AF693937

WITHDRAWN

Sea Mammals and Oil

Sea Mammals and Oil: Confronting the Risks

Edited by

Joseph R. Geraci and **David J. St. Aubin**
Department of Pathology
Ontario Veterinary College
University of Guelph
Guelph, Ontario, Canada

Academic Press, Inc.
Harcourt Brace Jovanovich, Publishers
San Diego New York Boston
London Sydney Tokyo Toronto

QL
713.2
.S43
1990

This book is printed on acid-free paper. ∞

Copyright © 1990 by Academic Press, Inc.
All Rights Reserved.
No part of this publication may be reproduced or transmitted in any form or by any means, electronic or mechanical, including photocopy, recording, or any information storage and retrieval system, without permission in writing from the publisher.

Academic Press, Inc.
San Diego, California 92101

United Kingdom Edition published by
Academic Press Limited
24–28 Oval Road, London NW1 7DX

Library of Congress Cataloging-in-Publication Data

Geraci, J. R.
Sea mammals and oil : confronting the risks / Joseph R. Geraci, David J. St. Aubin.
p. cm.
Includes bibliographical references.
ISBN 0-12-280600-X (alk. paper)
1. Marine mammals--Effect of oil spills on. 2. Oil spills--Environmental aspects. I. St. Aubin, D. J. II. Title.
QL713.2.G47 1990
599.5'045222--dc20 90-31984
CIP

Printed in the United States of America
90 91 92 93 9 8 7 6 5 4 3 2 1

Contents

Contributors

Numbers in parentheses indicate the pages on which the authors' contributions begin.

Joseph R. Geraci, (167, 211)
Department of Pathology, Ontario Veterinary College, University of Guelph, Guelph, Ontario N1G 2W1, Canada

Valerie Lounsbury, (241)
Department of Pathology, Ontario Veterinary College, University of Guelph, Guelph, Ontario N1G 2W1, Canada

Ian A. McLaren, (55)
Biology Department, Dalhousie University, Halifax, Nova Scotia B3H 4J1, Canada

Jerry M. Neff, (1, 35)
Battelle Ocean Sciences, Duxbury, Massachusetts 02332

Katherine Ralls, (199)
Department of Zoological Research, National Zoo, Smithsonian Institution, Washington, District of Columbia 20008

Donald B. Siniff, (199)
Department of Ecology and Behavioral Biology, University of Minnesota, Minneapolis, Minnesota 55455

David J. St. Aubin, (103, 235, 241)
Department of Pathology, Ontario Veterinary College, University of Guelph, Guelph, Ontario N1G 2W1, Canada

Ian Stirling, (223)
Canadian Wildlife Service, Edmonton, Alberta T6H 3S5, and Zoology Department, University of Alberta, Edmonton, Alberta T6G 2E9, Canada

Thomas D. Williams, (211)
Monterey Bay Acquarium, Monterey, California 93940

Bernd Würsig, (129)
Moss Landing Marine Laboratory, Moss Landing, California 95039-0450

Foreword

Here, ten specialists in petroleum chemistry, marine mammalogy, and veterinary medicine explore the effects of spilled petroleum (or "oil") on five groups of marine mammals: seals, whales and dolphins, sea otters, polar bears, and manatees. The authors hope to enlighten not only scientists and petroleum technologists but all those who labor in government, academia, or industry to preserve the vitality of the sea. Oil floating on the sea, spreading over its beaches, and penetrating deeply into beach sediments is patently offensive to human sensibilities. The authors ask: To what extent is oil also health-damaging to the sea mammals who must rub against it, or swim through it, or ingest it, or breathe its vapors?

This question leads to others. When mammals living at the site of an oil spill are clearly in distress, might not they have been in or near distress before the spill happened? What are the chronic (residual) effects, in addition to the acute (immediate) effects, of oil on marine mammals? Is due attention being given to the effects of oil flowing from urban and industrial sources? Collectively, these flows contribute more petroleum hydrocarbons than do spills from tankers and wells.

In this book, as in any useful scientific book, questions asked but not answered are almost as valuable as those answered. Mystery stimulates research. Did, or did not, the blowout of a Union Oil well off Santa Barbara in 1969 cause the premature births of over two hundred California sea lions? See pages 112–113.

The authors focus on the marine mammal species of North American waters, especially the coastal waters where oil spills most often occur. They offer an estimate of the vulnerability to oil of each species. The estimate is based on the known biology and ecology of the species; on the known effects of oil on the species; and on the *probable* effects, insofar as these can be predicted from study of the literature on oil damage to humans, domestic animals, and laboratory animals.

Although the effects of oil on marine bacteria, phytoplankton, invertebrates, fishes, and birds are fairly well known, the effects on marine mammals are not, for three main reasons.

- Marine mammals, while sharing certain adaptations to life in the sea—such as the possession of limbs modified as flippers and prowess in capturing subaquatic foods—are not closely related. They comprise five distinct zoological groups. Thus their life cycles, patterns of habitat use, and (as will be explained) responses to oil are different.
- Many of them live in remote regions where they must be studied with the help of costly *matériel* such as ships, aircraft, diving gear, and underwater electronic devices. Many are large-bodied, difficult to handle and to maintain in captivity. Their respiratory systems are strongly modified for diving (an elephant seal can hold its breath for over an hour). This peculiarity challenges the researcher who would anaesthetize or immobilize a marine mammal for study.
- Most marine mammals are protected by laws that shield them wherever they may be—swimming free or even stranded—from direct approach by researches. Although research *per se* is not banned, it is commonly impeded by the legal requirement to obtain a scientific permit. And in the end, those who would study marine mammals by invasive methods face growing widespread public resistance to "experimenting" on animals.

The oil spill issue is heating up. It is being fueled by rapid growth in the size and complexity of the oil industry and by a newly risen public concern for the permanence of planet Earth's natural environments. Call this a social conscience. The demand for oil which began scarcely a hundred years ago has outpaced the safety engineering, regulatory legislation, and corporate accountability that might have reduced the incidence of major spills. In one brief century, the United States (for example) consumed over 60% of the nation's total known and presumed, though untapped, oil deposits (Kerr, 1989; Goodrich, 1989). Meanwhile, year after year, the seagoing tankers of the world gave way to larger ones. What the new ones gained in economy, they lost in seaworthiness. Witness the fact that eight tankers, each spilling 100,000 tons or more of oil, ruptured at sea between 1967 and 1988 (Ware, 1989). Many other tankers spilled lesser amounts. The all-time world record spill of 600,000 tons flowed in 1979 from an offshore well, Ixtoc No. 1, in the Gulf of Mexico (*World Almanac*, 1989). Spills continue to enter the pages of history under the names of the ships or platforms implicated.

The first hundred thousand-ton spill helped to induce the birth of environmentalism, a body of principles and practices which soon found a permanent and necessary place in Western thought. Environmentalism gained momentum with the celebrations of Earth Day in 1970 and with the convening of the great United Nations Conference on the Human Environment in 1972. Environmentalism, broadly defined, is a movement toward understanding humankind's natural bases of support while continuously applying what is learned toward perpetuating those bases. Environmentalists—the champions of ennvironmentalism—often speak of posterity. They are, in a sense, among posterity's constituency. They speak, for example, of "temporal imperialism," meaning the arrogance with which living persons destroy, or allow to be destroyed, wildlife diversity—thereby impoverishing all future persons whose lives would otherwise have been enriched by the existence of that diversity. In our own time, oil spills could easily wipe out forever the California sea otters and the Mediterranean and Hawaiian monk seals. Perilously endangered, these three populations now comprise fewer than four thousand individuals.

The image of sea mammals trapped in oil is very real to the many millions who know little about the ocean and its animals but simply empathize with wildlife. Perhaps they have watched nature shows on television or have visited oceanariums. They react emotionally to oil spills—and always will. "We are," says the Irish storyteller Sean O' Faolain (1983), "for a great part of our lives at the mercy of uncharted currents of the heart."

But this book is less concerned with feelings than with risks. Civilization advances by risk assessment, entailing social cost-accounting and in turn the weighing of one good (such as oil) against another (such as sea otters). Although social cost-accounting is ultimately the job of administrators, legislators, and jurists, it must begin with, and rest on, facts that only scientists can provide. Thus the main thrust of the book is to give facts and educated guesses about the hazards to sea mammals posed by spilled oil. Although the authors do not admit to educated guessing, they do admit to state-of-the-art modeling. For instance, according to one model, an oil spill reaching 1,600 sea otters in Monterey Bay might kill 18.6% (or 340 animals). According to another model, oil reaching the fur seal rookeries on St. Paul Island might kill 10,603 to 73, 948 animals. And according to a third, "there is a 51.8% chance that at least one bowhead whale will encounter oil during the 30 to 40 year lifespan of the [Beaufort Sea] oil field." If the book serves its purpose:

- Planners in the oil industry will, where risk to marine mammals is high, pay more attention than in the past to the selection of drilling sites and transportation routes and to the integrity of carriers. Risk is always highest where species are rare or endangered or are especially vulnerable to oil damage (witness the sea otters).
- Planners in wildlife management will, where risk to marine mammals is high, insist on precautionary measures for regulating the production and transport of oil. They will also call for baseline studies of the ecosystem at risk. These studies will be valuable in the event that before-oil and after-oil comparisons of the ecosystem should be called for.
- Marine mammal rescue-and-rehabilitation teams on the site of a spill, aware that certain species are more vulnerable to damage than others, will be better able to apply triage.

Readers who approach this book with a bias for or against the oceanic production and transport of oil will find reinforcement of their held opinions. Readers who approach it with scientific detachment and willingness to learn will appreciate its fair treatment.

Victor B. Scheffer
Bellevue, Washington

Kerr, R. A. (1989). Oil and gas estimates plummet. *Science* **245**, 1330–1331.
Goodrich, L. J. (1989). US mulls shrinking oil supply. *Christian Science Monitor* **82**, 1–2.
"World Almanac and Book of Facts" (1989) p. 513.
Ware, L. (1989). Oil in the sea: The big spills and blowouts. *Audubon* **91**, 109.
O'Faolain, S. (1983). "Collected Stories" p. 584. New York: Atlantic.

Preface

In 1977, Robert Hofman, Scientific Program Director for the U.S. Marine Mammal Commission, saw the need to bring together a review of the effects of oil on marine mammals. The early 1970's embargo on oil from the Middle East had spurred a new era of offshore oil exploration and development, one that might threaten marine mammal populations that were already under pressure. We accepted the assignment, partly because we had earlier undertaken experimental studies on the effects of oil on seals, and because it provided an opportunity to bear down on the literature, to include accounts from newspapers, periodicals, and other popular sources that we normally tend to overlook.

The information available then was scarcely enough to answer a few questions, so the exercise, later published in Marine Fisheries Review, could easily point out lines of investigation designed to answer basic questions on how oil might be perceived by a sea otter or affect the peculiar skin of a whale or dolphin. The years since have witnessed the completion of several studies to answer such questions, and have also seen a number of oil spills—the largest in North America occurred in Valdez, Alaska, while this book was being written—that placed the experimental studies into perspective. Now, thirteen years later, it was time to bring together this new level of understanding.

To do this, we needed to delve beyond an evaluation of direct toxic effects and into the features of an animal's life history and habitat use that increase or reduce the chance that it will encounter oil. Here too, there have been many advances. Generous funding by government agencies has made possible long-term studies of natural history. Technology now provides the means to identify and track groups of cetaceans, follow the movements of seals, otters, and bears, and record the depths to which an animal might venture to gather food. This information has been brought together here by Ian McLaren for pinnipeds, Bernd Würsig for cetaceans, Katherine Ralls and Donald Siniff for sea otters, and Ian Stirling for polar bears. These authors have drawn on years of experience to examine when and how each group or species might face or interact with oil, thereby preparing the reader for a better understanding of the toxic effects.

The book opens with accounts by Jerry Neff on the nature and fate of oil in the marine environment and on some recent attempts to use computer models to predict the impacts of oil on populations as a whole. Following an introductory chapter on the life history and ecology of each marine mammal group is an analysis of the effects of oil. For those chapters we take sole responsibility, but drew in Thomas Williams for his firsthand experience rehabilitating oiled sea otters in Alaska, and Valerie Lounsbury for her eclectic understanding of the literature on manatees.

We are grateful for the help of many others: Valerie for the illustrations that complement the text and for bringing together information on all aspects of the subject, Dale-Marie Smith for capable typing and skillful handling of the manuscript, Lenore Latta for careful examination of the page proofs, Elizabeth Crossey for her work on the index, and Ted Eaton for preparing the photographs. Philip E. K. Symons applied his rigorous and incisive editorial skills, anad what we all missed did not escape Victor Scheffer's keen eye. While at the Battelle Memorial Institute, Eiji Imamura was instrumental in securing financial support for this project, and helped coordinate the efforts of the contributors. The project would not have been undertaken without the financial support of the Minerals Management Service (MMS), which funded our earlier research on the effects of oil on cetaceans, initiated the call for a review under the guidance of Robert Miller, and, spearheaded by the efforts of William Lang, supported the transformation and expansion of that monograph into publishable form.

Joseph R. Geraci
David J. St. Aubin

1

Composition and Fate of Petroleum and Spill-Treating Agents in the Marine Environment

Jerry M. Neff
Battelle Ocean Sciences
Duxbury, Massachusetts 02332

Introduction

The chemical properties of oil ultimately determine its effects. Some compounds are actively toxic, and damage delicate tissues, such as eyes, nasal cavities, and other sensitive mucous membranes. Their noxious properties are balanced by the fact that they are rapidly dissipated and removed from the environment. At the other end of the spectrum are the persistent forms, such as tar and weathered oil, which are not as toxic but have greater potential for environmental impact because of their resistance to weathering. To evaluate the consequences of oil exposure in marine mammals, it is important to understand these properties of petroleum: its composition, how it enters the marine environment, and what happens to it then.

Composition and Toxicity of Petroleum

Crude petroleum, a complex mobile mixture of fossil, biogenic origin (Speers and Whitehead, 1969), contains literally thousands of organic and a few inorganic compounds. Included within the classification are natural gas, liquid petroleum oils, resins, and asphaltenes. A more precise definition of petroleum is impossible, because no two are identical. Most crude

Sea Mammals and Oil: Confronting the Risks
Copyright © 1990 by Academic Press, Inc.
All rights of reproduction in any form reserved.

petroleums contain the same classes of compounds, but they differ in the relative amounts of each constituent.

Crude petroleum may contain organic compounds ranging in molecular weight from methane to complex polymeric structures such as asphaltenes with molecular weights of 100,000 or more (Kallio, 1976). Natural gas is separated from liquid petroleum at the time of production; the oil is then distilled to produce commercial products. Each fraction is collected at a different distillation temperature (Fig. 1–1), and can be refined further into a product with more desirable properties. The residue after distillation contains much of the resin and asphaltene fractions of the crude oil. It is a thick tarry liquid or solid that may be used for fuel (Bunker C residual oil, Number 6 fuel oil) or paving (asphalt).

Hydrocarbons (compounds composed only of carbon and hydrogen atoms) are the most abundant components of crude and refined petroleum (Fig. 1–2). They account for more than 90% of natural gas and from 50 to about 98% of liquid crude petroleum (Kallio, 1976; National Academy of Sciences, 1985). Other components include sulphur, oxygen, nitrogen, and a variety of metallic elements which are bound to organic compounds or exist as inorganic salts.

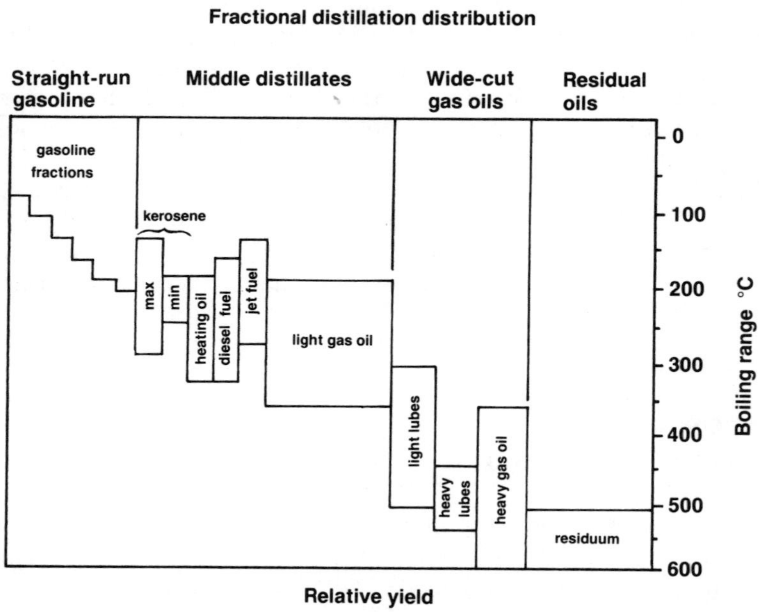

Figure 1-1
Boiling point range of fractions of crude petroleum. The width of each rectangle is proportional to the amount of the substance (National Academy of Sciences, 1985).

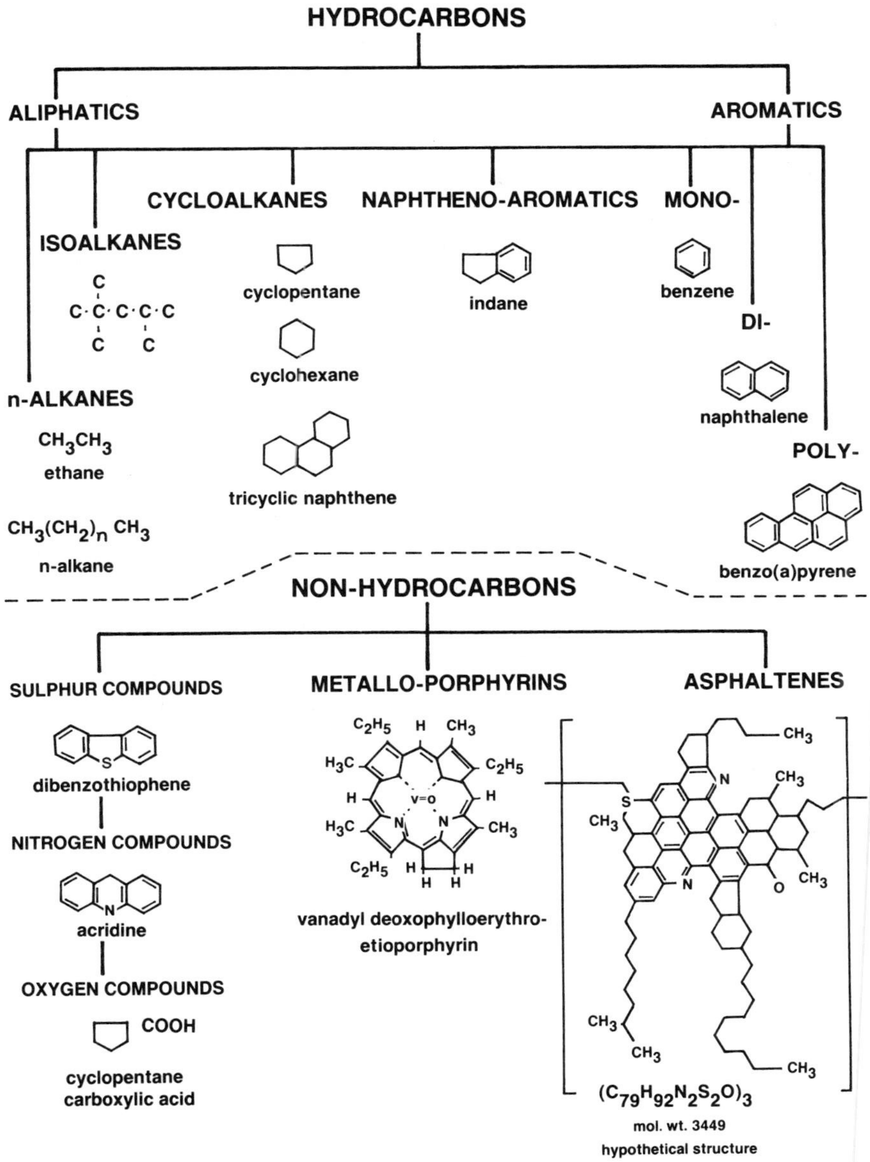

Figure 1-2
Examples of the chemical structure of some common components of crude petroleum (from Miller and Connell, 1982).

Alkanes, which contain single chemical bonds between carbon atoms, are the most abundant hydrocarbons in petroleum. There are three types: normal, branched, and cyclic. The normal and branched alkanes are usually present in about equal amounts. The n-alkanes range in size from methane (C_1) to about C_{60}, and possibly as high as C_{78}. A majority of the branched or isoalkanes are simple 2-, 3-, and 4-methylalkanes. They also include isoprenoid compounds that range from C_{10} to about C_{40}. The most abundant isoprenoid alkanes in petroleum are pristane (C_{19}) and phytane (C_{20}).

Cycloalkanes, also called cycloparaffins or naphthenes, may comprise as much as 50% of the total hydrocarbons in oil. Most are cyclopentane derivatives. They may contain aromatic ring structures, normal or branched alkane substituents, or nonhydrocarbon groups, such as one or more carboxylic acid moieties.

Aromatic hydrocarbons may account for about 20% of the total hydrocarbons in crude oil. The basic building block of an aromatic hydrocarbon is the benzene ring, a six-member carbon ring containing nine equally shared carbon–carbon covalent bonds. Benzene occurs in small amounts in natural gas, crude, and particularly the lighter fractions of refined oil. It may be linked to another benzene ring through a single carbon–carbon bond to form biphenyl. Polycyclic aromatic hydrocarbons (PAHs) are composed of two to nine or more fused benzene rings (Neff, 1979). Naphthalene ($C_{10}H_8$), which consists of two fused rings, is the lowest molecular weight PAH. The abundance of aromatic hydrocarbons in petroleum usually decreases with increasing molecular weight. In most cases, the one-ring (benzene) through three-ring (phenanthrene) compounds account for at least 90% of the aromatic hydrocarbons (Neff, 1979). These may combine with cycloalkanes to form naphtheno–aromatic compounds, and combine further to produce polymeric structures that are important components of the resin and asphaltene fractions of petroleum (Fig. 1-2).

The resin and asphaltene fractions of crude oil have not been well characterized (Speers and Whitehead, 1969; Kallio, 1976). They presumably consist of high molecular weight hydrocarbons and heterocompounds containing sulphur, oxygen, or nitrogen, and thermally induced condensation products of lower molecular weight aromatics and heteroaromatics. Asphaltenes are thought to be present in colloidal suspension.

Refined Oil Products

Refined petroleum products contain all the chemical classes present in crude oil, but primarily those compounds boiling over a fairly narrow temperature range (Fig. 1–1). For example, gasoline contains primarily low-boiling alkanes (C_5 to C_9) and monoaromatics, whereas residual oil contains more of high-boiling alkanes and polycyclic aromatic hydrocarbons, as well as most of the resins and asphaltenes originally present in the crude oil. In addition, catalytic cracking of the gasoline fraction produces a group of unsaturated compounds, the alkenes and cycloalkenes, that are ordinarily present only in trace concentrations in the original oil. The refining process may also increase the degree of alkylation of the alkane/alkene fraction. These changes improve the properties of gasoline as a fuel.

Similar Compounds of Natural and Pyrogenic Origin

Combustion of organic material, including fossil fuels, is a major source of PAHs containing three or more aromatic rings. Their formation is favored particularly if combustion takes place in an oxygen-deficient environment. Resulting PAH assemblages are complex and, unlike those in petroleum, are dominated by four-, five-, and six-ring aromatics. These differences are useful in distinguishing between petrogenic and pyrogenic hydrocarbon assemblages in environmental samples.

Toxicity of Petroleum Compounds

The composition of crude or refined petroleum governs its behavior and fate when spilled at sea. It also affects the responses of marine organisms, including mammals, that might contact the spilled oil.

Acute toxicity of alkanes to aquatic organisms tends to increase with molecular weight. However, acutely toxic concentrations for all but the lowest molecular weight alkanes are higher than their solubility, and therefore cannot occur naturally in aquatic environments (Hutchinson *et al.*, 1980). Low molecular weight cyclic alkanes (cyclohexane and several alkyl cyclohexanes) appear to be more toxic to aquatic organisms than n-alkanes and benzenes of similar molecular weight (Benville *et al.*, 1985). Mixtures of higher molecular weight alkanes, such as paraffin oils, are considered inert. In fact, they are used by humans as laxatives. Low molecular weight alkanes (methane through octane) have mild anesthetic properties (Crisp *et al.*, 1967), and, because of their volatility, may occur in a form that can be inhaled.

Aromatic hydrocarbons are the most toxic of the major classes of compounds in petroleum. The acute toxicity of crude and refined petroleums to aquatic organisms (Neff and Anderson, 1981; National Academy of Sciences, 1985) and mammals (U.S. Environmental Protection Agency, 1981) correlates directly with the concentration of light aromatic hydrocarbons (benzene through phenanthrene). Chronic effects of petroleum are attributed primarily to four- and five-ring aromatic and heteroaromatic hydrocarbons, some of which are well-known carcinogens (Karcher *et al.*, 1981; Oesch, 1982; Grunbauer and Wegener, 1983; Later *et al.*, 1983). Benzene, though a known carcinogen (Fishbein, 1984), is volatile and short-lived, and probably contributes more to acute than chronic toxicity.

The acute toxicity of aromatic hydrocarbons is inversely proportional to molecular weight (Neff, 1979; Hutchinson *et al.*, 1980). However, because of their low solubility, aromatic hydrocarbons with four or more rings rarely exist in acutely toxic concentrations. At the other extreme, the monocyclic aromatic hydrocarbons (benzene, toluene, and xylene) are so volatile that they are lost rapidly from water. Thus, naphthalenes and phenanthrenes, which are slightly soluble and relatively persistent, contribute most to the toxicity of crude and refined petroleum (Neff, 1979). Heterocyclic compounds can have a toxicity similar to the analogous aromatic hydrocarbons (Thomas *et al.*, 1981). Dibenzothiophene and several of its alkyl homologs are abundant in many crude oils, and therefore probably contribute to the toxicity of the oils.

Sources of Petroleum in the Marine Environment

On average, 10,000–15,000 metric tons of petroleum enter the marine environment each day. Miller and Connell (1982) estimated that, of the 3100 million metric tons of oil produced in 1981, from 4.5 to 6.1 million metric tons (0.15 to 0.20% of production) reached the oceans. The National Academy of Sciences (1985) gives a value of 3.2 million metric tons (more than 750 million gal) of oil entering the ocean per year (Table 1–1).

A wide variety of natural and anthropogenic sources contribute to this load (Table 1–1). Natural sources are the most difficult to estimate, but are currently thought to be on the order of 200,000 metric tons (National Academy of Sciences, 1985); one-fourth of this is eroded from oil-bearing rocks. Wilson *et al.* (1974) compiled a list of 190 known submarine oil seeps, and several more have been identified since. In U.S. outer continental shelf waters, there are 54 seeps off southern California, 28 off the south coast of Alaska, 3 along the coast of the Gulf of Mexico, and others scattered along the Bering and Beaufort Sea coasts of Alaska. They also occur in Mexican waters of the Gulf of Mexico, along the Caribbean coast of South America, and the northeast coast of Canada. Seeps discharge at different rates, with as much as 30,000 tons each year released in the Santa Barbara Channel alone (Fischer, 1978).

Table 1-1
Input of Petroleum Hydrocarbons into the Marine Environment in Millions of Metric Tons/Year[a]

Source	Probable range	Best estimate[b]
Natural sources		
Marine seeps	0.02–2.0	0.2
Sediment erosion	0.005–0.5	0.05
Total natural sources	0.025–2.5	0.25
Offshore production	0.04–0.06	0.05
Transportation		
Tanker operations	0.4–1.5	0.7
Dry-docking	0.02–0.05	0.03
Marine terminals	0.01–0.03	0.02
Bilge and fuel oils	0.2–0.6	0.3
Tanker accidents	0.3–0.4	0.4
Nontanker accidents	0.02–0.04	0.02
Total transportation	0.95–2.62	1.47
Atmosphere	0.05–0.5	0.3
Municipal and industrial wastes and runoff		
Municipal wastes	0.4–1.5	0.7
Refineries	0.06–0.6	0.1
Nonrefining industrial wastes	0.1–0.3	0.2
Urban runoff	0.01–0.2	0.12
River runoff	0.01–0.5	0.04
Ocean dumping	0.005–0.2	0.02
Total wastes and runoff	0.585–3.12	1.18
Total	1.7–8.8	3.2

[a] From National Academy of Sciences, 1985.
[b] The total best estimate, 3.2 million metric tons annually, is a sum of the individual best estimates. A value of 0.3 was used for the atmospheric inputs to obtain the total, although we well realize that this best estimate is only a center point between the range limits and cannot be supported rigorously by the data and calculation used for estimation of this input.

The most important anthropogenic source is that associated with marine transportation (Table 1–1). It is matched by the combined contribution from municipal and industrial wastes, offshore oil production activities, atmospheric deposition, runoff from rivers and urban areas, and ocean dumping, primarily of sewage sludge and industrial wastes (National Academy of Sciences, 1985). In any given year, there are over 10,000 incidents involving the spillage of oil in U.S. waters alone (U.S. Coast Guard, 1987).

Tanker operations and accidents account for most of the oil entering the ocean from marine transportation activities. Nothing attracts attention like a tanker spill. It is messy and harmful, but not always in direct proportion to its size. One of the largest spills occurred in 1978 when the *Amoco Cadiz* ran aground releasing approximately 60 million gal of crude oil along 350 km of the Brittany coast of France. Yet its effect on marine mammals was minimal compared to the 11-million-gal spill from the March 1989 grounding of the *Exxon Valdez* in Prince William Sound, Alaska. Spills and discharges from other vessels contribute only a small percentage of the oil released in any year, but may have a significant local impact on marine mammals. Specific events are reviewed in each of the ensuing chapters on toxic effects.

As unsightly as they are, tanker accidents are not the principal source of oil releases arising from marine transportation. More significant are discharges of ballast and tank-washing water. Such activities are regulated by the International Maritime Organization which allows discharge of oil from cargo areas of a tanker under way in international waters at a rate of no more than 60 L per mile, not to exceed 1/15,000 of the total cargo of older tankers and 1/30,000 of the cargo of new tankers. No such discharges are allowed in U.S. territorial waters or other designated areas, such as the Red and Mediterranean seas.

When a tanker arrives in ballast at an oil terminal, the water in dedicated ballast tanks is not contaminated with oil and can be discharged to local waters. However, when ballast water is carried in the cargo tanks, as in supertankers and old tankers, it may be discharged to an onshore treatment facility. There the water is separated from the oil and discharged. The treatment facility in Valdez, Alaska, during its first two years of operation, discharged to Valdez Harbor a total of 33.4 billion L of treated ballast water containing about 130 metric tons of particulate oil and 170 metric tons of volatile hydrocarbons (mainly benzenes) (Lysyj *et al.*, 1981).

Offshore oil exploration and production is viewed as a major source of spilled oil, and indeed it can be. The Ixtoc-I blowout in the Bay of Campeche in the Mexican Gulf of Mexico was the worst recorded oil spill of any kind. On June 3, 1979, an exploratory well about 80 km northwest of Cuidad de Carmen blew out. By the time the well was capped 290 days later on March 23, 1980, about 475,000 metric tons of oil had been lost (Jernelov and Linden, 1981). Usually, platform spills are of a much smaller magnitude, on the order of 40,000–60,000 metric tons per year (National Academy of Sciences, 1985). In waters of the U.S. continental shelf, the performance record for offshore platforms has been quite good (Minerals Management Service, 1986). Of the 690 million metric tons of oil produced there between 1970 and 1985, about 8400 metric tons, or 0.001% of production, were spilled.

Two types of discharges from offshore exploration and production platforms may contain oil: drilling muds and produced water. Drilling muds are mixtures of clays, weighting agents, and other ingredients in a water- or oil-base (National Academy of Sciences, 1983). They are used to lubricate the bit and offset pressure during the drilling of each well. In the North Sea and in Canadian waters, drill cuttings containing up to 10% diesel oil have been discharged. This could represent up to about 100 tons of oil discharged for each well drilled. Such practices are not permitted in U.S. waters where only water-base drilling muds can be discharged, but even they contain small amounts of oil (usually less than about 100 mg/kg).

Produced water is fossil water that emanates with the oil and gas from most wells. A well during its life yields approximately equal volumes of fossil fuel and produced water. The latter may be reinjected through another well to the reservoir, or treated to remove particulate oil and discharged. United States federal standards not yet promulgated (U.S. Environmental Protection Agency, 1985) would set a maximum allowable concentration of petroleum in produced water at 59 mg/L. The amount of produced water generated by a given well varies. A single production platform may discharge up to one million L, and a large treatment facility up to 10 million L or more each day. The National Academy of Sciences (1985) estimated that approximately 50 billion L of produced water are discharged to U.S. state and federal waters each year, carrying 1500 to 3000 metric tons of petroleum hydrocarbons.

Municipal and, to a lesser extent, industrial waste waters also deliver petroleum to the marine environment. The annual discharge of petroleum carried with sewage has been estimated to be about 13,000 tons in the Hudson–Raritan Estuary (Connell, 1982), 17,000 tons in the Southern California Bight (Eganhouse and Kaplan, 1982), and nearly 500 tons in central Puget Sound (Barrick, 1982). The National Academy of Sciences (1985) estimated that 200,000 metric tons of petroleum in municipal waste water were discharged each year to U.S. coastal waters. Industrial discharges, including those from oil refineries, contribute smaller quantities of petroleum. These and several other less notable sources probably constitute little potential hazard to marine mammals.

Fate of Petroleum in the Marine Environment

The timing and relative importance of physical, chemical, and biological processes affecting the fate of oil differ with each category of discharge and petroleum product. However, the types of processes are the same. The most important fall under the category of weathering, which includes spreading, evaporation, dissolution, dispersion into the water column, photochemical oxidation, formation of emulsions, microbial degradation, adsorption to suspended particulate matter, stranding on shore, or sedimentation to the sea floor (Payne and McNabb, 1985; Payne *et al.*, 1987; Boehm, 1987) (Figs. 1–3 and 1–4). Weathering changes the physical and

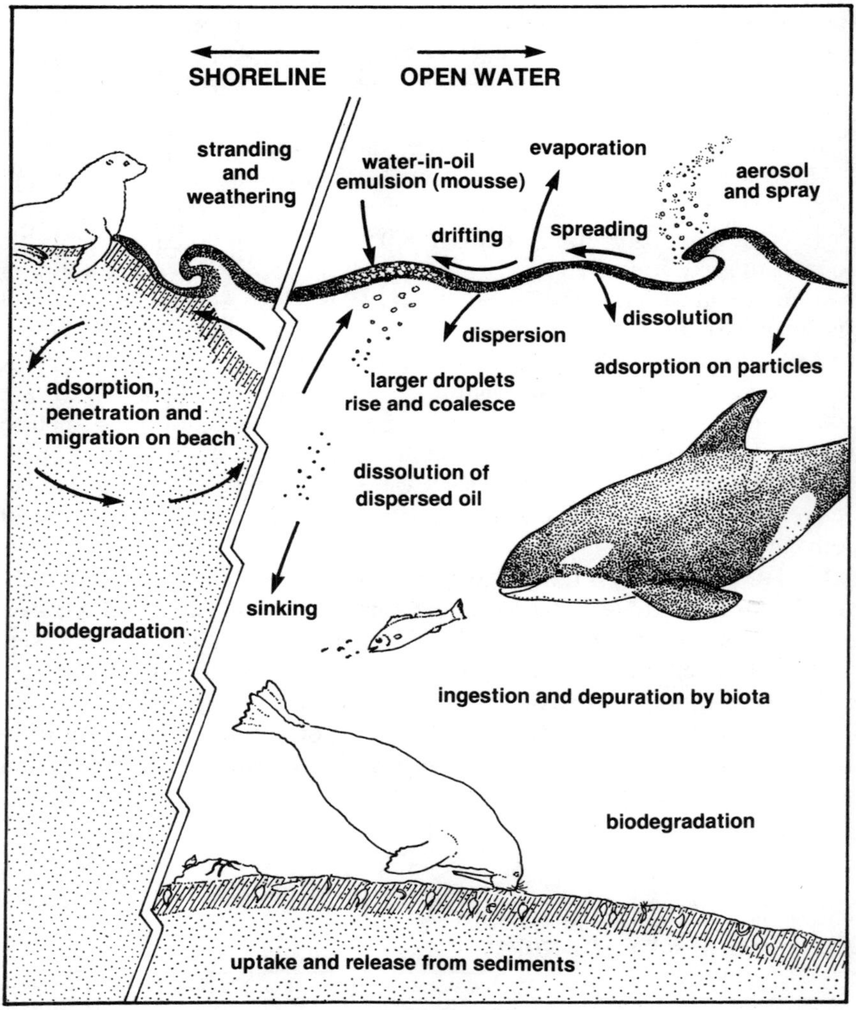

Figure 1-3
Behavior and fate of oil spilled in the marine environment (modified from Bobra and Fingas, 1986).

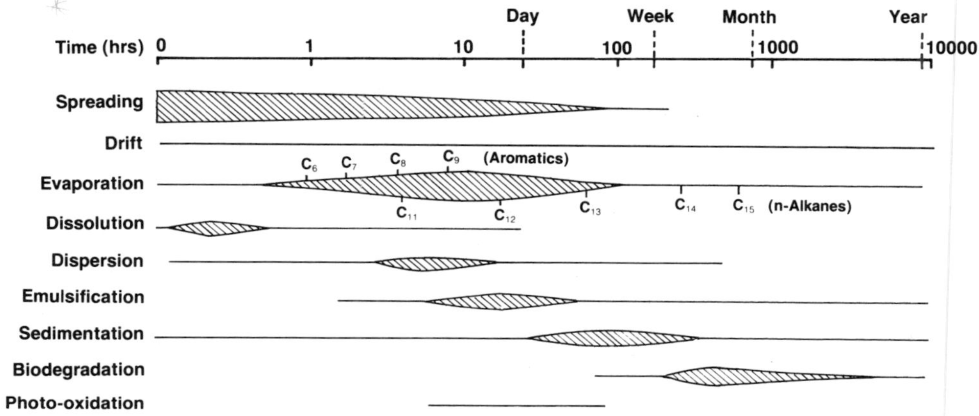

Figure 1-4

The timing of oil weathering processes following an oil spill on the sea surface. The length of the line indicates the probable timespan of a process, shown here on a logarithmic scale. The width of a line indicates the relative magnitude of the process through time and in relation to other concurrent weathering processes (from Wheeler, 1978).

chemical properties of spilled oil and thereby influences its toxicity to marine organisms.

Spreading and Drifting

Oil released at or near the sea surface will be affected first by spreading (Fig. 1–4). If discharged below the surface, it must rise through the water column before it can form an oil slick. Under such conditions, oil droplets form and disperse, and the lower molecular weight components dissolve (Boehm and Feist, 1982). Most of the petroleum discharged as part of a complex mixture such as municipal sewage never reaches the sea surface and is therefore not subjected to the same weathering forces as surface oil.

When oil is released on the sea surface, it spreads horizontally in an elongated pattern oriented in the direction of the wind and surface water currents (Elliott, 1986; Elliott *et al.*, 1986). The center of the mass of the slick may move at a rate of approximately three percent of the wind speed with a 20- to 30-degree shift to the right (in the northern hemisphere) due to the Coriolis effect (Payne and McNabb, 1985). Several mathematical models have been developed to predict the trajectories of oil slicks (Samuels *et al.*, 1983). The major axis of the slick tends to elongate at a linear rate with time, whereas the width of the slick grows as a function of $t^{0.5}$. Spreading is more rapid on warm than on cold water due to differences in viscosity of the oil; moderate wave action also increases the rate of spread. Crude oils and heavy distillates form two phases during spreading: a thick phase (1–20 mm thick), consisting of viscous, partly emulsified oil; and a thin sheen 0.01–0.001 mm thick (Audunson *et al.,* 1981). In addition, the leading edge of the slick tends to be thicker than the interior (Elliott, 1986). The thick oil usually forms small patches that subdivide as they continue to weather. These patches usually move downwind at a faster speed than the thinner slick, eventually leaving it behind.

The area of the ocean surface covered by an oil slick depends on the volume and composition of the oil, rate of discharge, and environmental conditions. An iridescent sheen is about 0.3–5 μm thick and has a concentration on the sea surface of about 1000 L/km^2 (National Academy of Sciences, 1985), whereas the surface concentration of a 1–10 mm-thick slick may be in the order of 10^6–10^7 L/km^2. Thus, the spill from the *Argo Merchant* on Nantucket Shoals, Massachusetts, in December 1976 (7.9×10^6 gal) first covered as a thick slick an area of only 2.9 km^2, and later, as a sheen, an area of 29,000 km^2. A reasonable average thickness for a crude oil slick undergoing moderate weathering would be 0.1–1.0 mm; such a spill would occupy 100–1000 m^2/metric ton of oil.

Evaporation

For the first few days after a spill, evaporation is the most important weathering process affecting the volume and composition of oil. The type of oil, surface area of the slick, and environmental conditions influence the rate of evaporation (Wheeler, 1978), which for any given substance is directly proportional to its vapor pressure (Mackay and Leinonen, 1975; Wheeler, 1978) and inversely proportional to molecular weight (Fig. 1–5). Aromatic hydrocarbons tend to evaporate more rapidly than alkanes of similar molecular weight (see Fig. 1-4), despite the lower vapor pressure of the former, apparently because aromatics have a higher activity coefficient than alkanes in the oil phase (Harrison *et al.,* 1975). Light distillate fractions, such as gasoline, kerosene, and jet fuel, may evaporate completely, and as much as 60% of light crude oil may evaporate within a week or so after a spill (Wheeler, 1978). Owing to the inverse relationship between temperature and vapor pressure, hydrocarbons evaporate more slowly in cold Arctic waters than in temperate or tropical waters (Reijnhart and Rose, 1982).

Evaporation profoundly affects physical and chemical properties of a slick. The loss of volatile components increases density and viscosity, and reduces vapor pressure and toxicity (Bobra and Fingas, 1986). Tarry resin and asphaltene fractions increase, promoting the formation of water-in-oil emulsions and tar balls. These in turn slow the rate of diffusion of

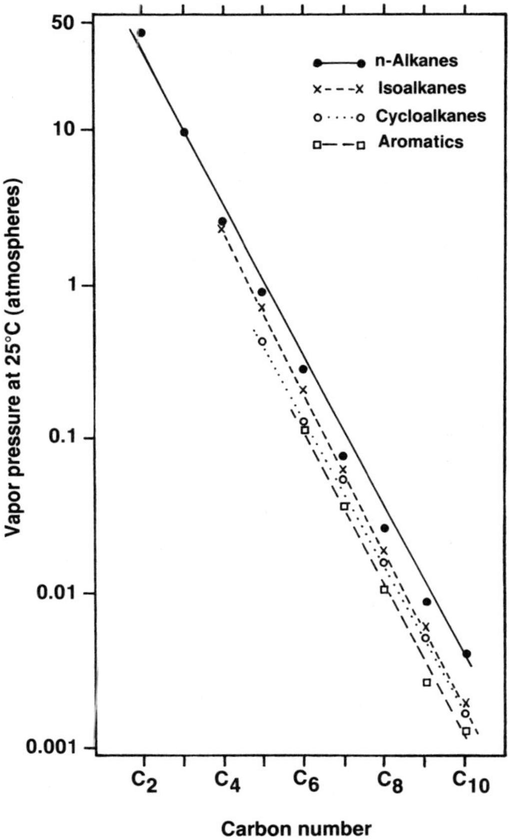

Figure 1-5

Relation between carbon number and vapor pressure of four classes of hydrocarbons in petroleum (from Wheeler, 1978).

remaining volatile hydrocarbons. Thus, emulsified oil, tar balls, and tar mats may develop a crust composed primarily of nonvolatile oil components covering a core of less weathered oil containing high concentrations of light hydrocarbons (Butler, 1975; Boehm and Feist, 1982; Payne and Phillips, 1985).

Dissolution

Dissolution into the water column, one of the first stages in the weathering process, removes less than 2–5% of a spill (Harrison *et al.*, 1975;

McAuliffe, 1977; Payne *et al.,* 1987). The process may nonetheless be significant because it brings the most toxic hydrocarbons into contact with marine organisms in a form that is readily available. In the Ixtoc-I blowout, lighter hydrocarbons partitioned into the water as the oil rose from the sea floor; benzene under the slick reached concentrations greater than 100 μg/L (Payne *et al.,* 1983). Dissolved and dispersed hydrocarbons persisted in the water column for up to 40 km from the blowout site (Boehm and Feist, 1982). Low concentrations of volatile aromatic hydrocarbons have also been detected in the water 2–3 km from the treated ballast water discharge at Port Valdez, Alaska (Lysyj *et al.,* 1981).

Dispersion and Emulsion Formation

When a slick is subjected to turbulent mixing or wave action, the larger droplets tend to coalesce, rise rapidly, and concentrate near the surface (Forrester, 1971). Small droplets break away from the main mass and become dispersed in the water column. If the droplets are small enough (less than 0.1 mm), they rise so slowly as to remain dispersed indefinitely (Payne and McNabb, 1985).

Dispersion is the most important process in the breakup and disappearance of a slick already reduced by evaporation. Dispersion begins soon after oil is discharged, reaches a peak within 10 hours (see Fig. 1–4), and within 100 hours overtakes spreading as the primary mechanism of transport of oil from the spill site (Wheeler, 1978). The activity and effectiveness of the process is due in part to the viscosity of the oil (Gordon *et al.,* 1983) and to the presence of natural surfactants (Wheeler, 1978) which facilitate droplet formation and inhibit coalescence.

Some oils, particularly after weathering, accumulate and retain water droplets within the oil phase (Mackay, 1982). These stable water-in-oil emulsions, sometimes called "chocolate mousse," may contain up to 75% water, and are more viscous than the parent oil. The concentration of heavy resin and asphaltene materials as well as endogenous surfactants determine the tendency of oil to form mousse (Payne and Phillips, 1985). Such emulsions form readily in the presence of sea ice (Payne and Phillips, 1985) and tend to collect beneath it.

The formation of stable water-in-oil emulsions slows the subsequent weathering of oil by inhibiting evaporation and dissolution of light fractions and photochemical and microbial degradation of the heavier fractions. Whether the slick forms a water-in-oil emulsion or an oil-in-water emulsion appears to depend on the viscosity, thickness, and chemical composition of the oil (Mackay, 1982), and on environmental factors.

Photochemical Reactions

Solar radiation acting on oil in the water generates photochemical reactions which yield new, mostly polar, compounds. Even in low concentrations (Ducreux *et al.*, 1986) these substances affect toxicity and behavior of the spilled oil (Payne and Phillips, 1985). The primary mechanism of photodegradation is photo-oxygenation (Larson *et al.*, 1976, 1977; Thominette and Verdu, 1984a), which yields peroxides, aldehydes, ketones, alcohols, and fatty acids (Payne and McNabb, 1985) that tend to be more water-soluble and toxic than the unoxidized parent compounds (Larson *et al.*, 1979). The process also yields high molecular weight byproducts that are not soluble in oil or water (Thominette and Verdu, 1984b).

Direct photolytic reactions, not requiring molecular oxygen, are quantitatively the most important mechanism of light-induced transformation (Zepp and Schlotzhauer, 1979; Mill *et al.*, 1981). For PAHs, the tendency toward direct photolysis increases with increasing molecular weight. For example, the half-life of naphthalene (two rings) in surface fresh water in sunlight at 40° N latitude in midsummer is 71 hours, compared to a half-life of 8 hours for phenanthrene (three rings), and 0.54 hours for benzo(a)pyrene (five rings).

Light intensity decreases with water depth, and so does the rate of photolysis of aromatic hydrocarbons. Photolysis is also inhibited at high latitudes where there is less sunlight, especially during winter (Fig. 1–6). At 60° N latitude, there is an approximately tenfold decrease in the rate of photolysis of benzo(a)pyrene between June and December (Zepp and Baughman, 1978). Photolytic rates of some compounds, such as benzo(b)thiophene and carbazole, are more sensitive to light intensity than others such as benz(a)anthracene and benzo(a)pyrene (Mill *et al.*, 1981).

Biodegradation

Marine bacteria and fungi play an important role in degrading and removing petroleum hydrocarbons from surface slicks, the water column, and sediments. Microbial degradation begins a day or so after the spill and continues as long as hydrocarbons persist (Wheeler, 1978; Lee and Ryan, 1983). Rate of degradation is related to oxygen concentration, temperature, nutrients (especially nitrogen and phosphorus), salinity, physical state and chemical composition of the spilled oil, and previous history of oil pollution at the spill site (Atlas, 1981; Bartha and Atlas, 1987).

Following a spill, all hydrocarbon components are degraded simultaneously, but at widely different rates, by indigenous water column and sediment microbiota (Atlas *et al.*, 1981, Bartha and Atlas, 1987). Low molecular weight n-alkanes in the C_{10} to C_{22} chain length range are

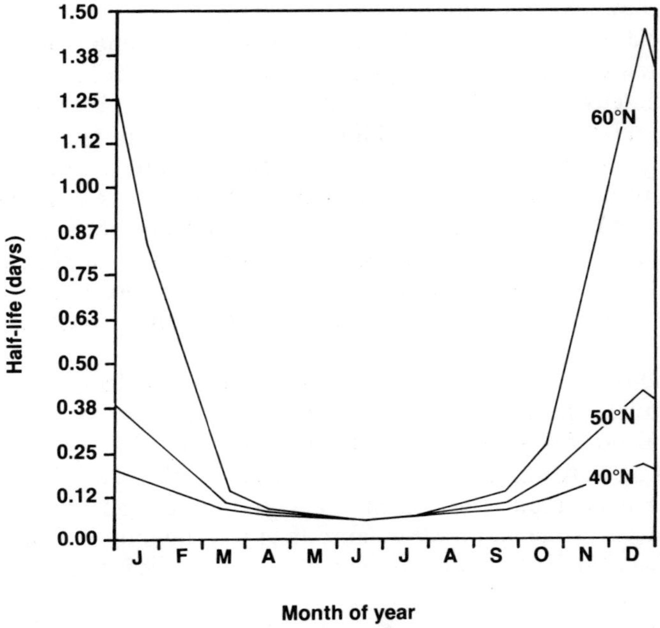

Figure 1-6
Annual variation in half-life (t_{50}) of benzo(a)pyrene dissolved in near-surface water at northern latitudes (from Zepp and Baughman, 1978).

metabolized more rapidly, followed by isoalkanes and higher molecular weight n-alkanes, olefins, monoaromatics, PAHs, and finally, highly condensed cycloalkanes, resins, and asphaltenes. Thus, as oil weathers through a combination of physical, photochemical, and biodegradative processes, it loses low molecular weight components and becomes enriched in higher molecular weight, more complex saturates, naphthenoaromatics, PAHs, resins, and asphaltenes.

Temperature profoundly affects the process of degradation. The half-life for microbial degradation of phenanthrene at an initial concentration of 25 μg/L in seawater is 79 days at 18°C and 11,000 days at 2°C. Similarly, that for benz(a)anthracene at an initial concentration of 2.5 mg/kg in sediment is 1100 days at 15°C and 21,000 days at 4°C (Lee and Ryan, 1983). The reliance on temperature was underscored by Wakeham *et al.* (1985, 1986), who showed that in summer conditions biodegradation was more important than volatilization in removing toluene, octadecane, and decane from the water column: under winter conditions, their contributions were reversed. Because both processes are diminished at low environmental temperatures, the light fractions of crude and refined petroleum are persistent in Arctic environments, especially in winter when low light intensity inhibits photo-oxidation.

Biodegradation is at best a slow process. Rates for hydrocarbons have been estimated to be 1–10 mg/m^3/day in open-ocean waters (Butler *et al.*, 1976), 30 mg/m^3/day beneath the surface slick produced by the *Amoco Cadiz* spill (Aminot, 1981), and 50 mg/m^2/day in the upper 5 cm of intertidal sediments along the Brittany coast of France fouled by the *Amoco Cadiz* spill. Extrapolating to the length of coastline affected, Atlas and Bronner (1981) estimated that it would take more than 20 years to biodegrade the estimated 64,000 tons on the Brittany coast. Yet before that time, the processes that combine to degrade the oil will have long removed it from view.

In Arctic environments, biodegradation is slower still, limited by lack of nutrients (Bergstein and Vestal, 1978; Atlas, 1986) and low temperatures (Cundell and Traxler, 1973; Gibbs *et al.*, 1975). Nevertheless hydrocarbon-degrading bacteria abound (Atlas, 1986) and can be coaxed into activity by exposure to petroleum (Button and Robertson, 1985).

Deposition in Sediments

Heavier fractions of oil eventually deposit in bottom sediments and persist for a long time. Sedimentation may occur by (1) adsorption of droplets on suspended (including biological) particles, and transport with them to the bottom; (2) stranding or beaching of oil, followed by adsorption onto sediments or erosion of hardened oil from substrates and subsequent transport to subtidal sediments; and (3) direct sinking of heavy or weathered oils (Anderson *et al.*, 1986; Boehm, 1987).

Suspended particles interact with spilled oil in two ways. They physically collide and adhere to dispersed droplets, and adsorb and partition dissolved hydrocarbons from the water phase (Payne *et al.*, 1987). A key variable in adsorption appears to be the concentration of suspended particulate matter, especially clay, in the water column. The greater the suspended sediment load, the more oil may be adsorbed and transported to the bottom (Boehm, 1987). Approximately 120–300 mg of petroleum may adsorb to each kilogram of suspended clay (Bassin and Ichiye, 1977; Meyers and Oas, 1978).

Weathered oil may become heavier than seawater and sink (Boehm, 1987). The process is enhanced as the density of water is lowered by influx of freshwater as runoff or from melting ice. In areas of significant downwelling, as in a polynya at the edge of an ice sheet, sinking water may carry oil droplets to the bottom. Additional oil may be fixed onto biological particles, particularly zooplankton fecal pellets.

Beached oil can also contribute to the load in subtidal sediments. Erosion of the beach by seasonal storms or ice-scouring results in transport of oil-laden sand into the subtidal zone. Studies of the Baffin Island experimental oil spill (Boehm *et al.*, 1985) and the *Amoco Cadiz* oil spill (Gundlach *et al.*, 1983) have shown that concentrations in excess of 100 ppm oil can deposit in subtidal sediments if oil comes ashore and subsequently erodes from the beach.

Outcome of a Spill

The physical impact and fate of a spill will depend on the type of oil, location of the event, and climatic conditions. Estuaries and salt marshes are particularly sensitive because they trap oil and thereby delay or prevent it from weathering. The Arctic too is sensitive; low temperatures retard the breakdown of oil and prolong recovery. Other environments more readily recover.

As one example, Gundlach *et al.* (1983) described the fate of the 60 million gal of oil spilled from the *Amoco Cadiz*. More than a third of it came ashore in bays and estuaries and contaminated anoxic intertidal and subtidal sediments. Approximately 30% of the oil evaporated, and about 5% was degraded in the water column by bacteria. More than 20% of the oil was unaccounted for and probably was carried away as surface slicks or tar balls. The most persistent oil residues were those incorporated into subtidal sediments of estuaries or nearshore waters, or buried in shifting sands. Most of the oil had disappeared from the water within a few months after the spill.

Oil Dispersants

Between 1967 and 1979, chemical dispersants were used to combat at least 16 major oil spills. These substances transform the oil into fine droplets that disperse in the water column and can be carried away and diluted by normal ocean mixing processes. Most dispersants contain surface active agents (surfactants), a solvent, and stabilizing agents (Tetra Tech, Inc., 1985; Canevari, 1986). A surfactant contains both a hydrophilic (water-compatible) and a hydrophobic (oil-compatible) group which allows it to concentrate at the boundary between oil and water (Fig. 1–7). The effect is to reduce surface tension, and thereby promote greater dispersion. There are three types of surfactants categorized according to the nature of the hydrophilic group: anionic, cationic, and nonionic. Nonionic surfactants are used most frequently and include ethoxylated alkylphenols such as nonyl phenol-ethylene oxide, ethoxylated linear alcohols such as oleyl

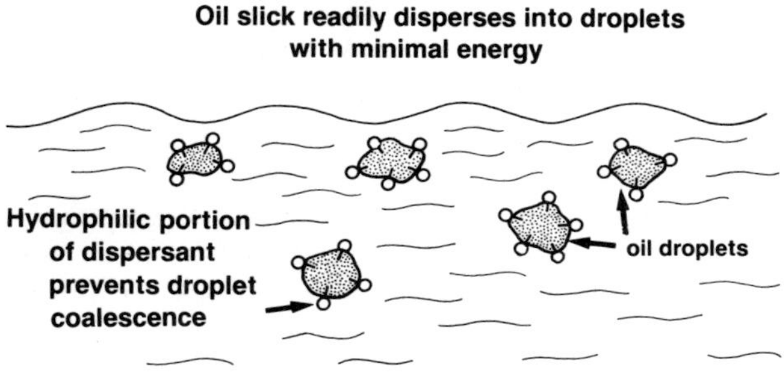

Figure 1-7
Mechanism of action of surfactants in modern oil dispersants (from Canevari, 1986).

alcohol, and esters formed by the reaction of fatty acids with polyhydric alcohols.

Solvents are added to lower the freezing point and reduce viscosity, and so make the dispersant easier to apply. The most common are aliphatic and aromatic hydrocarbon mixtures, water or alcohols in water, glycols, and glycol ethers. Dispersants containing hydrocarbon solvents tend to be more effective in treating heavy, viscous oils. They are easier to mix and apply, but are more toxic than those containing water-soluble solvents. A stabilizer may also be added to adjust pH, reduce corrosiveness, help fix the dispersion after it is formed, or counteract adverse color or odor. Dispersant stabilizers may include alkalis, phosphates, silicates, nitrates, dyes, and polymerized alkyl naphthalene sulfonates.

Dispersants may be applied to an oil slick by hand or from a suitably equipped boat or aircraft (Fickling and Hann, 1980). Small to medium-sized boats with spray booms, usually about 7–10 m long extending out from each side, are used most frequently. They apply water-based preparations at a rate of about 100–150 gal per minute; concentrates and hydrocarbon-based agents are applied at a lower rate. Aerial spraying has the advantage of covering large areas quickly. Dispersants are applied full-strength from aircraft flying 30–50 ft above the surface.

The volume of dispersant required to treat a spill depends on sea state, nature of the oil, and method of application. Generally, the rougher the seas and the fresher the oil (less weathered), the less dispersant required to break up the slick. The ratio of dispersant to oil ranges from 1 : 10 to 1 : 1. During the Ixtoc-I spill, several dispersants were used, most of which were applied by plane. Up to four spraying missions were flown per day, each applying about 36,000 gallons. Additional dispersant was applied from boats. Clearly an unacceptably large amount of dispersant is required to treat a major spill.

Dispersants have also been used to loosen oil on shore so that it can be removed more easily by cleaning devices or by wave action. This practice was discouraged after the *Torrey Canyon* oil spill in 1967 on the southwest coast of Cornwall, England, because it caused more damage to coastal ecosystems than did the oil itself (Nelson-Smith, 1968; Southward and Southward, 1978). Less toxic dispersants for shoreline cleaning are now being tested, but even with the use of less toxic dispersants the advantages are questionable. Studies on rocky shores and intertidal mud flats reveal little difference in impact between raw crude oil and crude oil that was treated with a dispersant after stranding (Little *et al.*, 1981; Rowland *et al.*, 1981; Crothers, 1983). Results from comparable studies in a salt marsh are somewhat more encouraging (Baker *et al.*, 1980).

Fate of Dispersants

Chemical dispersants applied to oil spills undergo the same weathering processes as the oil (Tetra Tech, Inc., 1985). These include evaporation, dissolution, diffusion, biodegradation, and possibly bioaccumulation (Wells *et al.*, 1982). Much of the hydrocarbon or water-soluble solvent fraction evaporates during and immediately after the dispersant is applied, especially when applied as a fine spray from an airplane.

When applied to a surface spill, the dispersant immediately partitions between the oil and water phases. Mackay and Hossain (1982) estimated that the oil–water partition coefficient for surfactants most frequently used in oil dispersants is about 10. Because the ratio of oil to water in nearly all spills is very low, most of the dispersant (up to about 99%) partitions into the water phase. Once in aqueous solution, the dispersant is diluted by diffusion and convective mixing, but surfactant components are detectable in concentrations of 1–3 ppm for over 6 hours after application to an oil spill (Bocard *et al.*, 1984).

The complex fatty acid ester mixtures usually used as surfactants in modern oil dispersants are readily degraded by marine bacteria and fungi. Several types of marine water column and sediment bacteria are capable of rapid and sustained growth with oil dispersant as the sole source of carbon and energy (Liu, 1983). More than 55% of the dispersant BP1100X was degraded in 8 days by a mixed population of microbes isolated from oil-contaminated sediments (Bhosle and Row, 1983). Microbial degradation probably is the most important mechanism quantitatively for removing dispersants from the marine environment.

A variety of aquatic organisms readily accumulate (Kikuchi *et al.*, 1980) and metabolize (Payne, 1982) surfactants from oil dispersants. Enzymatic hydrolysis of the surfactant yields hydrophilic and hydrophobic components. The former probably are excreted via the gills and kidneys, whereas the latter accumulate in the gallbladders of fish and are excreted very slowly. Metabolism of surfactants is rapid enough that there is little likelihood of food chain transfer from marine invertebrates and fish to consumers, including marine mammals.

Other Methods of Treating Spills

A number of other chemicals have been proposed for treating marine oil spills. These include herding agents, demulsifiers, and gelling, wicking, and sinking agents. Most of these are in the conceptual or developmental stage, and their chemical compositions are proprietary.

Herding agents have a higher surface tension, and therefore spreading force, than petroleum (Fickling and Hann, 1980). When applied around a slick, they tend to compress it and prevent it from spreading. They reportedly have relatively low toxicity to marine organisms. Demulsifiers are mixtures of surfactants and wetting agents intended to facilitate the separation of the oil and water in emulsions (Canevari, 1982). The environmental properties and toxicity of demulsifiers should be similar to those of dispersants. Gelling agents transform spilled oil into a semisolid mass that can be handled easily with recovery equipment. By injecting them into the oil in cargo or fuel tanks of a sunken ship, they may slow or prevent release of oil. These compounds apparently have low toxicity to marine organisms (Tokuda, 1979).

Spilled oil can be effectively burned, particularly if fireproof booms are used (Buist *et al.*, 1983). However, water serves as a heat sink making it difficult to sustain combustion. Not all the oil burns, and the residues and airborne particles may create added problems. Wicking agents are intended to lift the oil above the sea surface, enabling it to burn more efficiently. They have not been used successfully in open water, but may prove more beneficial on ice when oil gathers in relatively deep pools (Mackay, 1982).

Sinking agents are dense particles (2.4–3.0 g/cm^3) with hydrophobic surfaces that, when applied to a spill, adsorb the oil and cause it to sink. Sand, fly ash, powdered cement, or other minerals coated with silicones, stearates, or waxes have been used. They effectively remove oil from sight, but they may exacerbate the impact of the spill by rapidly depositing the oil on the bottom where it may persist.

Interactions between Petroleum and Marine Mammals

In North American waters, approximately 95% of offshore production is in the Gulf of Mexico, especially off Louisiana. The other principal sites are off southern California, in Alaska at Cook Inlet, the Beaufort Sea from the North Slope of Alaska to the MacKenzie River delta, and the Grand Banks on the east coast of Canada. Tanker traffic is heaviest in the Gulf of Mexico along the Texas coast, along the Pacific coast from Alaska to southern California (the main tanker route for Alaskan Prudhoe Bay oil to refineries in Washington and California), and along the Atlantic coast, where vessels transport crude petroleum from abroad, and refined products to major urban markets. All of these areas are used by marine mammals, as displayed on distribution maps in later chapters. Based on these distributions, one might project that most encounters between marine mammals and oil spills would occur along the California coast, in the Gulf of Alaska, and on the Grand Banks of eastern Canada. In fact the area is broader than that; any vessel fueled by petroleum is a potential source of a spill.

The drilling platform *Kulluk* in the Beaufort Sea. Photo by B. Würsig.

Physical Contact with Oil

All marine mammals spend considerable time at the surface, swimming, breathing, feeding, or resting, thereby enhancing the possibility of contact with a surface slick, water-in-oil emulsion, or tar balls. In species with heavy pelage, such as fur seals, sea otters, and polar bears, contact may lead to fouling. Polar bears and otters groom themselves regularly as a means of maintaining the insulating properties of the fur, and may thereby ingest oil. Oil would have less tendency to adhere to the surface of animals with relatively little or no pelage, such as whales, dolphins, manatees, and most seals.

Some baleen whales forage at the surface, a behavior called skim-feeding (Würsig *et al.*, 1985). When feeding in an area of a slick or tar balls, they may foul the feeding apparatus. Tarry residues in particular could coat the baleen plates.

In polar regions, spilled oil tends to accumulate at the ice edge, in leads, polynyas, and breathing holes (Fig. 1–8), where animals such as narwhals, belugas, ringed seals, walruses, and polar bears spend much of their time. The oil tends to persist, thus setting the stage for unavoidable contact.

Oil that comes ashore is likely to foul pinnipeds that require such areas for haulouts or nursery areas, and, to a lesser extent, may foul otters

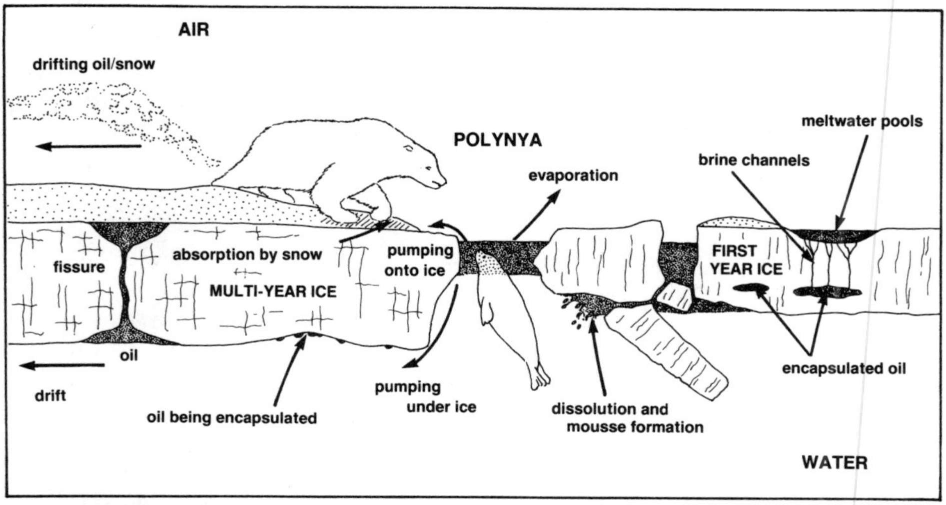

Figure 1-8
Behavior and fate of oil spilled on or around floating ice (modified from Bobra and Fingas, 1986).

and bears. Some of the oil is eventually returned in subtidal sediments, where it may transfer to species such as the gray whale, walrus, and some seals that feed heavily on benthic animals.

Accumulation of Oil from Air and Water

Marine mammals encountering fresh oil are likely to inhale volatile hydrocarbons evaporating from the surface slick. Such fractions contain toxic monoaromatic hydrocarbons (benzenes, toluenes, xylenes) and low molecular weight aliphatics with anesthetic properties. Inhalation of these compounds is potentially dangerous (Carpenter *et al.,* 1975, 1976).

Inhalation of concentrated petroleum vapors may cause inflammation of mucous membranes, lung congestion, or even pneumonia (Hansen, 1985). Volatile hydrocarbons, such as benzene and toluene, that are inhaled are transferred rapidly into the bloodstream from the lungs. They may accumulate from the blood in such tissues as brain and liver, causing neurological disorders and liver damage (Geraci and St. Aubin, 1982).

Marine mammals probably will not accumulate much oil directly from solution or dispersion in the water column. The skin of cetaceans seems relatively impermeable to oil (Geraci and St. Aubin, 1982). Most marine mammals do not drink large volumes of seawater, so significant accumulation of hydrocarbons by this route is unlikely.

Fur-bearing marine mammals such as fur seals, polar bears, and sea otters may ingest oil during grooming. The limited data available indicate that oil is not particularly toxic, at least to pinnipeds, when taken in by this route. Geraci and Smith (1977) showed that seals experienced no acute damage when they ingested 75 mL of oil over a short period of time. However, ingestion of oil during grooming may have contributed to the death of heavily oiled polar bears (Oritsland *et al.,* 1981) and sea otters (Geraci and Williams, this volume, Chapter 8).

Ingestion of Oil-Contaminated Food

Marine mammals, except the manatee, are carnivores that rely on invertebrates or fish for sustenance. Their feeding strategies could lead to ingestion of oil-contaminated food, because most of the prey organisms can accumulate petroleum hydrocarbons in their tissues (Neff, 1979; Capuzzo, 1987).

Zooplankton are a particularly important food resource, particularly for baleen whales. Copepods, euphausiids, and mysids become contaminated by assimilating hydrocarbons directly from seawater and by ingesting oil droplets and tainted food (Corner, 1978). Copepods are one of the few taxa in which hydrocarbon uptake appears to be easier from food than from water (Corner *et al.*, 1976). There is an inverse relationship between ambient temperature and rate of accumulation of petroleum hydrocarbons by copepods (Harris *et al.*, 1977); polar and boreal species store more lipids, and therefore hydrocarbons, than those from warm environments. Planktonic crustaceans can transform aromatic hydrocarbons to polar metabolites that may be excreted or bound to tissues (Malins, 1977). For a few days or weeks, unmetabolized or metabolized hydrocarbons in zooplankton could be transferred to predators (Corner *et al.*, 1976).

Benthic invertebrates may accumulate petroleum hydrocarbons from contaminated sediments and food and, to a greater extent, from water (Neff, 1979, 1984; Capuzzo, 1987). Bivalve molluscs, with limited ability to metabolize the compounds to excretable polar metabolites (Lee, 1981), accumulate higher concentrations and retain them longer than other taxa (Neff and Anderson, 1981; Capuzzo, 1987). By contrast, benthic crustaceans have a well-developed mixed-function oxidase (MFO) system to eliminate these compounds (Lee, 1981). Therefore, marine mammals such as the walrus (Oliver *et al.*, 1983) and sea otter that rely heavily on bivalve molluscs for food share a higher risk of ingesting petroleum hydrocarbons than do gray whales which forage on benthic crustaceans (Nerini and Oliver, 1983). The greater threat to gray whales is that benthic amphipods are quite sensitive to spilled oil; they are among the first marine animals killed and the populations are slow to recover (Spies, 1987). Thus, a major spill in the northern Bering Sea in summer could affect the whales' main food resource.

Marine fish also take up petroleum hydrocarbons from water and food. The compounds induce the hepatic MFO system (Stegeman, 1981), and within a few days after exposure, aromatic hydrocarbons are oxygenated to polar metabolites and excreted. For this reason, most fish do not accumulate and retain high concentrations of petroleum hydrocarbons, even in heavily oil-contaminated environments, and so are not likely to transfer them to predators. Fish may nevertheless be tainted with metabolites bound to tissue macromolecules, including DNA. The metabolites are so reactive, it is unlikely that they would be released in a toxic form during digestion by the consumer, and so would not pose a serious threat.

Marine carnivores generally are inefficient assimilators of petroleum compounds in food. For this reason, and because all prey species are able to release hydrocarbons from their tissues (Neff and Anderson, 1981), marine food chain biomagnification does not occur. Thus, there is no direct correlation between a marine mammal's trophic level and the concentration of residues that it might consume. In fact, top carnivores such as polar bears and killer whales that feed on large pelagic fish and seals are less likely to be exposed to petroleum in their food than are species such as baleen whales and walrus that feed on zooplankton and benthic invertebrates.

Effects of Oil Dispersants

Virtually nothing is known about the effects of oil dispersants on marine mammals, except as they are used to clean oil-fouled sea otters. By removing spilled oil from the sea surface, dispersants obviously reduce the risk of contact. The oil remaining, on the one hand, would be less sticky, and therefore less likely to adhere to fur, skin, baleen plates, or other body surfaces. On the other hand, the surfactants in dispersants may remove natural oils from marine mammal fur, thereby decreasing its insulating properties. Cleaning oiled beaches and rocky shores with dispersants may be an effective means of preventing oiling of pinnipeds that may wish to haul out there. More work needs to be done before we can adequately weigh the advantages or disadvantages of using dispersants in such habitats.

Conclusion

Representatives of all major groups of marine mammals can be found in North American waters in habitats ranging from tropical rivers to the polar ice pack. To determine how any one species might be threatened by oil requires an understanding of its life history and ecology as presented in the following chapters.

References

Aminot, A. (1981). Anomalies du système hydrobiologique cotier apres l'échouage de l'Amoco Cadiz. Considérations qualitatives et quantitatives sur la biodégradation in situdes hydrocarbures. *In* "*Amoco Cadiz:* Conséquences d'une Pollution Accidentelle par les Hydrocarbures (Fates and Effects of the Oil Spill)," pp. 223–242. CNEXO, Paris.

Anderson, J. W., Neff, J. M., and Boehm, P. D. (1986). "Sources, Fates and Effects of Aromatic Hydrocarbons in the Alaskan Marine Environment with Recommendations for Monitoring Strategies," EPA 600/3-86 018. Environmental Protection Agency, Corvallis, Oregon.

Atlas, R. M. (1981). Microbial degradation of petroleum hydrocarbons. An environmental perspective. *Microbiol. Rev.* **45,** 180–209.

Atlas, R. M. (1986). Fate of petroleum pollutants in Arctic ecosystems. *Water Sci. Technol.* **18,** 59–67.

Atlas, R. M., and Bronner, A. (1981). Microbial hydrocarbon degradation within intertidal zones impacted by the *Amoco Cadiz* oil spillage. *In* "*Amoco Cadiz:* Conséquences d'une Pollution Accidentelle par les Hydrocarbures (Fates and Effects of the Oil Spill)," pp. 251–256. CNEXO, Paris.

Atlas, R. M., Boehm, P. D., and Calder, J. A. (1981). Chemical and biological weathering of oil from the *Amoco Cadiz* spillage within the littoral zone. *Estuarine, Coastal Shelf Sci.* **12,** 589–608.

Audunson, V. D., Mathisen, J., Holdorsen, J., and Krough, K. (1981). Slikforkast— a simulation program for oil spill emergency tracking and long term contingency planning. *In*" PETROMAR80—Petroleum in the Marine Environment," pp. 453–459. Association Europeenne Oceanique EUROCEAN. Graham & Trotman Ltd., London.

Baker, J. M., Crothers, J. H., and Mullett, J. A. J. (1980). Ecological effects of dispersed and non-dispersed crude oil: A progress report. *In* "Petroleum Development and the Environment, 1979 Annual Conference Proceedings," pp. 85–100. Institute of Petroleum, London.

Barrick, R. C. (1982). Flux of aliphatic and polycyclic aromatic hydrocarbons to central Puget Sound from Seattle (Westpoint) primary sewage effluent. *Environ. Sci. Technol.* **16,** 682–692.

Bartha, R., and Atlas, R. M. (1987). Transport and transformations of petroleum: Biological processes. *In* "Long-Term Environmental Effects of Offshore Oil and Gas Development" (D. F. Boesch and N. N. Rabalais, eds.), pp. 287–341. Elsevier Appl. Sci., London.

Bassin, J. J., and Ichiye, T. (1977). Flocculation behavior of suspended sediments and oil emulsions. *J. Sediment Petrol.* **47,** 671–677.

Benville, P. E., Jr., Whipple, J. A., and Eldridge, M. B. (1985). Acute toxicity of seven acyclic hexanes to striped bass, *Morone saxatilis,* and bay shrimp, *Crangon franciscorum,* in seawater. *Calif. Fish Game* **11,** 132–140.

Bergstein, P. E., and Vestal, J. R. (1978). Crude oil biodegradation in Arctic tundra ponds. *Arctic* **31**(3), 158–169.

Bhosle, N. B., and Row, A. (1983). Effect of dispersants on the growth of indigenous bacterial populations and biodegradation of oil. *Indian J. Mar. Sci.* **12,** 194–196.

Bobra, A. M., and Fingas, M. F. (1986). The behavior and fate of Arctic oil spills. *Water Sci. Technol.* **18,** 13–23.

Bocard, C., Castaing, G., and Gatillier, C. (1984). Chemical oil dispersion in trials at sea and in laboratory tests: The key role of dilution processes. *ASTM Spec. Tech. Publ.* **STP 840,** 125–142.

Boehm, P. D. (1987). Transport and transformation processes regarding hydrocarbon and metal pollutants in offshore sedimentary environments. *In* "Long-Term Environmental Effects of Offshore Oil and Gas Development" (D. F. Boesch and N. N. Rabalais, eds.), pp. 233–286. Elsevier Appl. Sci., London.

Boehm, P. D., and Feist, D. L. (1982). Subsurface distribution of petroleum from an offshore well blowout—the *Ixtoc-I* blowout. *Environ. Sci. Technol.* **16,** 67–74.

Boehm, P. D., Steinhauer, W., Requejo, A., Cobb, D., Duffy, S., and Brown, J. (1985). Comparative fate of chemically dispersed and untreated oil in the Arctic: The BIOS studies 1980–1983. *In* "Proceedings of the 1985 Oil Spill Conference." Am. Petrol. Inst., Washington, D.C.

Buist, I. A., Pistruzak, W. M., Potter, S. G., Vanderkooy, N., and McAllister, I. R. (1983). The development and testing of a fireproof boom. *In* "Proceedings of the 1983 Oil Spill Conference; Prevention, Behavior, Control, Cleanup," pp. 43–51. Am. Petrol. Inst., Washington, D.C.

Butler, J. N. (1975). Evaporative weathering of petroleum residues: The age of pelagic tar. *Mar. Chem.* **3,** 9–21.

Butler, J. N., Morris, B. F., and Sleeter, T. D. (1976). The fate of petroleum in the open ocean. *In* "Sources, Effects and Sinks of Hydrocarbons in the Aquatic Environment," pp. 287–297. Am. Inst. Biol. Sci., Washington, D.C.

Button, D. K. and Robertson, B. R. (1985). Effect of toluene exposure time and concentration on induction of high affinity values for toluene oxidation by bacteria of estuarine seawater samples. *Mar. Ecol.: Prog. Ser.* **26,** 187–193.

Canevari, G. P. (1982). The formulation of an effective demulsifier for oil spill emulsions. *Mar. Pollut. Bull.* **13,** 49–54.

Canevari, G. P. (1986). Oil spill dispersants: Mechanism, history, chemistry. "Committee on Effectiveness of Oil Spill Dispersants in the Ocean," Position Paper. Nat. Res. Counc., Natl. Acad. Sci., Washington, D.C.

Capuzzo, J. M. (1987). Biological effects of petroleum hydrocarbons: Assessments from experimental results. *In* "Long-Term Environmental Effects of Offshore Oil and Gas Development" (D. F. Boesch and N. N. Rabalais, eds.), pp. 343–410. Elsevier Appl. Sci., London.

Carpenter, C. P., Kinkead, E. R., Geary, D. L., Sullivan, L. J., and King, J. M. (1975). Petroleum hydrocarbon toxicity studies. V. Animal and human responses to vapors of mixed xylenes. *Toxicol. Appl. Pharmacol.* **33,** 543–558.

Carpenter, C. P., Kinkead, E. R., Geary, D. L., Sullivan, L. J., and King, J. M. (1976). Petroleum hydrocarbon toxicity studies. XIII. Animal and human response to vapors of toluene concentrate. *Toxicol. Appl. Pharmacol.* **36,** 473–490.

Connell, D. W. (1982). An approximate petroleum hydrocarbon budget for the Hudson-Raritan Estuary—New York. *Mar. Pollut. Bull.* **13,** 89–93.

Corner, E. D. S. (1978). Pollution studies with marine plankton. Part I. Petroleum hydrocarbons and related compounds. *Adv. Mar. Biol.* **15,** 289–380.

Corner, E. D. S., Harris, R. P., Kilvington, C. C., and O'Hara, S. C. M. (1976). Petroleum compounds in the marine food web: Short-term experiments on the fate of naphthalene in *Calanus. J. Mar. Biol. Assoc. U.K.* **56,** 121–133.

Crisp, D. J., Christie, A. O., and Ghobashy, A. F. A. (1967). Narcotic and toxic action of organic compounds to barnacle larvae. *Comp. Biochem. Physiol.* **22,** 629–649.

Crothers, J. H. (1983). Field experiments on the effects of crude oil and dispersant on the common animals and plants of rocky sea shores. *Mar. Environ. Res.* **8,** 215–239.

Cundell, A. M., and Traxler, R. W. (1973). Microbial degradation of petroleum at low temperature. *Mar. Pollut. Bull.* **4,** 125–127.

Ducreux, J., Berthou, F., and Bodennec, G. (1986). Etude duviellissement d'un petrole brut repandu à la surface de l'eau de mer dans des conditions naturelles. *Int. J. Environ. Anal. Chem.* **24,** 85–111.

Eganhouse, R. P., and Kaplan, I. R. (1982). Extractable organic matter in municipal wastewaters. I. Petroleum hydrocarbons: Temporal variations and mass emission rates to the ocean. *Environ. Sci. Technol.* **16,** 180–186.

Elliott, A. J. (1986). Shear diffusion and the spread of oil in the surface layers of the North Sea. *Dtsch. Hydrogr. Z.* **39,** 113–137.

Elliott, A. J., Hurford, N., and Penn, C. J. (1986). Shear diffusion and the spreading of oil slicks. *Mar. Pollut. Bull.* **17,** 308–313.

Fickling, J. M., Jr., and Hann, R. W., Jr. (1980). "The Chemical Treatment of Oil Spills: Theoretical, Practical, and Philosophical Considerations," Unpublished report. Texas A&M University, College Station.

Fischer, P. J. (1978). Natural gas and oil seeps, Santa Barbara Basin. *In* "The State Lands Commission 1977, California Gas, Oil, and Tar Seeps," pp. 1–62. Sacramento, California.

Fishbein, L. (1984). An overview of environmental and toxicological aspects of aromatic hydrocarbons. I. Benzene. *Sci. Total Environ.* **40,** 189–218.

Forrester, W. D. (1971). Distribution of suspended oil particles following the grounding of the tanker Arrow. *J. Mar. Res.* **29,** 151–170.

Geraci, J. R., and St. Aubin, D. J., (1982). "Study of the Effects of Oil on Cetaceans." U.S. Dept. of the Interior, Minerals Manage. Serv., Washington, D.C.

Geraci, J. R., and Smith, T. G. (1977). Consequences of oil fouling on marine mammals. *In* "Effects of Petroleum on Arctic and Subarctic Marine Environments and Organisms" (D.C. Malins, ed.), Vol. 2, pp. 399–410. Academic Press, New York.

Gibbs, C. F., Pugh, K. B., and Andrews, A. R. (1975). Quantitative studies on marine biodegradation of oil. II. Effect of temperature. *Proc. R. Soc. London, Ser. B* **188,** 83–94.

Gordon, D. C., Jr., Keizer, P. D., and Prouse, N. J. (1983). Laboratory studies on the accommodation of some crude and residual fuel oils in sea water. *J. Fish. Res. Board Can.* **40,** 1611–1618.

Grunbauer, H. J. M., and Wegener, J. W. M. (1983). The relation between chemical structure and mutagenic activity of some polycyclic aromatic sulfur heterocycles (S-PAH). *Toxicol. Environ. Chem.* **6,** 225–239.

Gundlach, E. R., Boehm, P. D., Marchand, M., Atlas, R. M., Ward, D. M., and Wolfe, D. A. (1983). Fate of *Amoco Cadiz* oil. *Science* **221,** 122–129.

Hansen, D. J. (1985). "The Potential Effects of Oil Spills and Other Chemical Pollutants on Marine Mammals Occurring in Alaskan Waters." OCS Rep. MMS85-0031. U.S. Dept. of the Interior, Minerals Manage. Serv., Alaska Outer Continental Shelf Region, Anchorage.

Harris, R. P., Berdugo, V., Corner, E. D. S., Kilvington, C. C., and O'Hara, S. C. M. (1977). Factors affecting the retention of petroleum hydrocarbons by marine planktonic copepods. *In* "Fate and Effects of Petroleum Hydrocarbons in Marine Organisms and Ecosystems" (D. A. Wolfe, ed.), pp. 286–304. Pergamon, New York.

Harrison, W., Winnik, M. A., Kwong, P. T. Y., and Mackay, D. (1975). Crude oil spills. Disappearance of aromatic and aliphatic components from small sea-surface slicks. *Environ. Sci. Technol.* **9,** 231–234.

Hutchinson, T. C., Hellebust, J. A., Tam, D., Mackay, D., Mascarenkas, R. A., and Shiu, W. Y. (1980). The correlation of the toxicity to algae of hydrocarbons and halogenated hydrocarbons with their physical-chemical properties. *In* "Hydrocarbons and Halogenated Hydrocarbons in the Aquatic Environment" (B. K. Afghan and D. Mackay, eds.), pp. 577–586. Plenum, New York.

Jernelov, A., and Linden, O. (1981). *Ixtoc-I:* A case study of the world's largest oil spill. *Ambio* **10,** 299–306.

Kallio, R. E. (1976). The variety of petroleums and their degradations. *In* "Sources, Effects and Sinks of Hydrocarbons in the Aquatic Environment," pp. 214–223. Am. Inst. Biol. Sci., Washington, D.C.

Karcher, W., Nelson, A., Depaus, R., van Eijk, J., Glaude, P., and Jacob, J. (1981). New results in the detection, identification, and mutagenic testing of heterocyclic polycyclic aromatic hydrocarbons. *In* "Chemical Analysis and Biological Fate: Polynuclear Aromatic Hydrocarbons" (M. Cooke and A. J. Dennis, eds.), pp. 317–328. Battelle Press, Columbus, Ohio.

Kikuchi, M., Wakabayashi, M., and Kojima, H. (1980). Bioaccumulation profiles of 35S-labelled sodium alkylpoly(oxyethylene) sulfates in carp (*Cyprinus carpio*). *Water Res.* **14,** 1541–1548.

Larson, R. A., Blankenship, D. W., and Hunt, L. L. (1976). Toxic hydroperoxides: Photochemical formation from petroleum constituents. *In* "Sources, Effects and Sinks of Hydrocarbons in the Aquatic Environment," pp. 298–308. Am. Inst. Biol. Sci., Washington, D.C.

Larson, R. A., Hunt, L. L., and Blankenship, D. W. (1977). Formation of toxic products from a No. 2 fuel oil by photooxidation. *Environ. Sci. Technol.* **11,** 492–496.

Larson, R. A., Bott, T. L., Hunt, L. L., and Rogenmuser, K. (1979). Photooxidation products of a fuel oil and their antimicrobial activity. *Environ. Sci. Technol.* **13,** 965–969.

Later, D. W., Pelroy, R. A., Mahlum, D. D., Wright, C. W., Lee, M. L., Weimer, W. C., and Wilson, B. W. (1983). Identification and comparative genotoxicity of polycyclic aromatic hydrocarbons and related nitrogen-containing heteroaromatic species in products from coal liquefaction processes. *In* "Polynuclear Aromatic Hydrocarbons: Formation, Metabolism and Measurement" (M. Cooke and A. J. Dennis, eds.), pp. 771–783. Battelle Press, Columbus, Ohio.

Lee, R. F. (1981). Mixed function oxygenases (MFO) in marine invertebrates. *Mar. Biol. Lett.* **2,** 87–105.

Lee, R. F., and Ryan, C. (1983). Microbial and photochemical degradation of polycyclic aromatic hydrocarbons in estuarine waters and sediments. *Can. J. Fish. Aquat. Sci.* **40,** 86–94.

Little, D., Baker, J. M., and Abiss, T. P. (1981). The fate and effects of dispersant-treated compared with untreated crude oil, with particular reference to sheltered intertidal sediments. *In* "Chemical Dispersion of Oil Spills: An International Research Symposium" (D. Mackay, ed.), Publ. No. EE-17, pp. 117–151. University of Toronto, Ontario, Canada.

Liu, D. (1983). Fate of oil dispersants in aquatic environments. *Sci. Total Environ.* **32,** 93–98.

Lysyj, I., Perkins, G., Farlow, J. S., Morris, R. W. (1981). Distribution of aromatic hydrocarbons in Port Valdez, Alaska. *In* "Proceedings of the 1981 Oil Spill Conference," pp. 47–54. Am. Petrol. Inst., Washington, D.C.

Mackay, D. (1982). Fate and behavior of oil spills. *In* "Oil and Dispersants in Canadian Seas—Research Appraisal and Recommendations" (J. B. Sprague, J. H. Vandermeulen, and P. G. Wells, eds.), pp. 7–27. Environ. Protect. Serv., Environ. Emergencies Branch, Ottawa, Ontario, Canada.

Mackay, D., and Hossain, K. (1982). Interfacial tensions of oil, water, chemical dispersant systems. *Can. J. Chem. Eng.* **60,** 546–550.

Mackay, D., and Leinonen, P. J. (1975). Rate of evaporation of low-solubility contaminants from water bodies to atmosphere. *Environ. Sci. Technol.* **9,** 1178–1180.

Malins, D. C. (1977). Biotransformation of petroleum hydrocarbons in marine organisms indigenous to the Arctic and sub-Arctic. *In* "Fate and Effects of Petroleum Hydrocarbons in Marine Organisms and Ecosystems" (D. A. Wolfe, ed.), pp. 47–59. Pergamon, New York.

McAuliffe, C. D. (1977). Evaporation and solution of C2 to C10 hydrocarbons from crude oils on the sea surface. *In* "Fate and Effects of Petroleum Hydrocarbons in Marine Organisms and Ecosystems" (D. A. Wolfe, ed.), pp. 19–35. Pergamon, New York.

Meyers, P. A., and Oas, T. G. (1978). Comparison of association of different hydrocarbons with clay particles in simulated seawater. *Environ. Sci. Technol.* **12,** 934–937.

Mill, T., Mabey, W. R., Lan, B. Y., and Baraze, A. (1981). Photolysis of polycyclic aromatic hydrocarbons in seawater. *Chemosphere* **10,** 1281–1290.

Miller, G. J., and Connell, D. W. (1982). Global production and fluxes of petroleum and recent hydrocarbons. *Int. J. Environ. Stud.* **19,** 273–280.

Minerals Management Service (1986). "Managing Oil and Gas Operations on the Outer Continental Shelf." U.S. Dept. of the Interior, Minerals Manage. Serv., Washington, D.C.

National Academy of Sciences (1983). "Drilling Discharges in the Marine Environment." Natl. Acad. Press, Washington, D.C.

National Academy of Sciences (1985). "Oil in the Sea." Natl. Acad. Press, Washington, D.C.

Neff, J. M. (1979). "Polycyclic Aromatic Hydrocarbons in the Aquatic Environment. Sources, Fates and Biological Effects." Appl. Sci. Publ., London.

Neff, J. M. (1984). Bioaccumulation of organic micropollutants from sediments and suspended particulates by aquatic animals. *Fresenius, Z. Anal. Chem.* **319,** 132–136.

Neff, J. M., and Anderson, J. W. (1981). "Response of Marine Animals to Petroleum and Specific Petroleum Hydrocarbons." Halstead Press, New York.

Nelson-Smith, A. (1968). The effects of oil pollution and emulsifier cleansing on shore life in south-west Britain. *J. Appl. Ecol.* **5,** 97–107.

Nerini, M. K., and Oliver, J. S. (1983). Gray whales and the structure of the Bering Sea benthos. *Oecologia* **59,** 224–225

Oesch, F. (1982). Chemical carcinogenesis by polycyclic aromatic hydrocarbons. *In* "Chemical Carcinogenesis" (C. Nicolini, ed.), pp. 1–24. Plenum, New York.

Oliver, J. S., Slattery, P. N., O'Connor, E. F., and Lowry, L. F. (1983). Walrus, *Odobenus rosmarus*, feeding in the Bering Sea: A benthic perspective. *Fish. Bull.* **81,** 501–512.

Øritsland, N. A., Engelhardt, F. R., Juck, F. A., Hurst, R. J., and Watts, P. D. (1981). "Effects of Crude Oil on Polar Bears," Northern Affairs Program Environ. Study No. 24. Dept. of Indian and Northern Affairs, Ottawa, Ontario, Canada.

Payne, J. F. (1982). Metabolism of complex mixtures of oil spill surfactant compounds by a representative teleost (*Salmo gairdneri,* crustacean (*Cancer irroratus*), and mollusc (*Chlamys islandicus*). *Bull. Environ. Contam. Toxicol.* **28,** 277–280.

Payne, J. R., and McNabb, G. D., Jr. (1985). Weathering of petroleum in the marine environment. *Mar. Technol. Soc. J.* **18,** 1–19.

Payne, J. R., and Phillips, C. R. (1985). "Petroleum Spills in the Marine Environment. The Chemistry and Formation of Water-in-Oil Emulsions and Tar Balls." Lewis Publishers, Chelsea, Michigan.

Payne, J. R., Kirstein, B. E., McNabb, G. D., Jr., Lambach, J. C., DeOliveira, C., Jordan, R. E., and Hom, W. (1983). Multivariate analysis of petroleum hydrocarbon weathering in the sub-Arctic marine environment. *In* "Proceedings of the 1983 Oil Spill Conference," pp. 423–434. Am. Petrol. Inst. Washington, D.C.

Payne, J. R., Phillips, C. R., and Hom, W. (1987). Transport and transformations: Water column processes. *In* "Long-term Environmental Effects of Offshore Oil and Gas Development" (D. F. Boesch and N. N. Rabalais, eds.), pp. 175–231. Elsevier Appl. Sci., London.

Reijnhart, R., and Rose, R. (1982). Evaporation of crude oil at sea. *Water Res.* **16,** 1319–1325.

Rowland, S. J., Tibbetts, P. J. C., and Little, D. (1981). The fate and effects of dispersant-treated compared with untreated crude oil, with particular reference to sheltered intertidal sediments. *In* "Proceedings of the 1981 Oil Spill Conference," pp. 283–293. Am. Petrol. Inst., Washington, D.C.

Samuels, W. B., LaBelle, R. B., and Amstutz, D. E. (1983). Applications of oil spill trajectory models to the Alaskan outer continental shelf. *Ocean Manage.* **8,** 233–250.

Southward, A. J., and Southward, E. C. (1978). Recolonization of rocky shores in Cornwall after use of toxic dispersants to clean up the *Torrey Canyon* spill. *J. Fish. Res. Board Can.* **35,** 682–706.

Speers, G. C., and Whitehead, E. V. (1969). Crude petroleum. *In* "Organic Geochemistry: Methods and Results" (G. Eglinton and M. R. J. Murphy, eds.), pp. 638–675. Springer-Verlag, Berlin.

Spies, R. B. (1987). The biological effects of petroleum hydrocarbons in the sea: Assessments from the field and microcosms. *In* "Long-Term Environmental Effects of Offshore Oil and Gas Development" (D. F. Boesch and N. N. Rabalais, eds.), pp. 411–467. Elsevier Appl. Sci., London.

Stegeman, J. J. (1981). Polynuclear aromatic hydrocarbons and their metabolism in the marine environment. *In* "Polycyclic Hydrocarbons and Cancer" (H. V. Gelboin and P. O. P. Ts'o, eds.), Vol. 3, pp. 1–60. Academic Press, New York.

Tetra Tech, Inc. (1985). "Fate and Effects of Oil Dispersants and Chemically Dispersed Oil in the Marine Environment," MMS-85-0048. U.S. Dept. of the Interior, Minerals Manage. Serv., Pacific Outer Continental Shelf Region, Los Angeles, California.

Thomas, P., Wofford, H. W., and Neff, J. M. (1981). Biochemical stress responses of striped mullet (*Mugil cephalus* L.) to fluorene analogs. *Aquat. Toxicol.* **1,** 329–342.

Thominette, F., and Verdu, J. (1984a). Photo-oxidative behavior of crude oils relative to sea pollution. Part I. Comparative study of various crude oils and model systems. *Mar. Chem.* **15,** 91–104.

Thominette, F., and Verdu, J. (1984b). Photo-oxidative behavior of crude oils relative to sea pollution. Part II. Photo-induced phase separation. *Mar. Chem.* **15,** 105–115.

Tokuda, H. (1979). Fundamental studies on the influence of oil pollution upon marine organisms. V. The toxicity of the gelling compounds for petroleum to marine organisms. *Bull. Jpn. Soc. Sci. Fish.* **45,** 1385–1387.

U.S. Coast Guard (1987). "Polluting Incidents in and around U.S. Waters," COMDTINST M16450.2G. U.S. Dept. of Transport., U.S. Coast Guard, Washington, D.C.

U.S. Environmental Protection Agency (1981). "Report to Congress on Listing of Waste Oil as a Hazardous Waste Pursuant to Section (8)(2)," Public Law 96-463. U.S. E.P.A., Washington, D.C.

U.S. Environmental Protection Agency (1985). "Development Document for Effluent Limitations Guidelines and Standards for the Offshore Segment of the Oil and Gas Extraction Point Source Category," EPA 440/1-85/055. U.S. E.P.A., Industrial Tech. Div., Washington, D.C.

Wakeham, S. G., Canuel, E. A., Doering, P. H., Hobbie, J. E., Helfrich, J. V. K., and Lough, G. R. G. (1985). The biogeochemistry of toluene in coastal seawater: Radiotracer experiments in controlled ecosystems. *Biogeochemistry* **1,** 307–328.

Wakeham, S. G., Canuel, E. A., and Doering, P. H. (1986). Behavior of aliphatic hydrocarbons in coastal seawater: Mesocosm experiments with [14C] octadecane and [14C] decane. *Environ. Sci. Technol.* **20,** 574–580.

Wells, P. G., Abernathy, S., and Mackay, D. (1982). Study of oil-water partitioning of a chemical dispersant using an acute bioassay with marine crustaceans. *Chemosphere* **11,** 1071–1086.

Wheeler, R. B. (1978). "The Fate of Petroleum in the Marine Environment," Spec. Rep. Exxon Production Research Co., Houston, Texas.

Wilson, R. D., Monaghan, P. H., Osanik, A., Price, L. C., and Rogers, M. A. (1974). Natural marine oil seepage. *Science* **184,** 857–865.

Würsig, B., Dorsey, E. M., Fraker, M. A., Payne, R. S., and Richardson,W. J. (1985). Behavior of bowhead whales, *Balaena mysticetus,* summering in the Beaufort Sea: A description. *Fish. Bull.* **83,** 357–377.

Zepp, R. G., and Baughman, G. L. (1978). Prediction of photochemical transformation of pollutants in the aquatic environment. *In* "Aquatic Pollutants; Transformation and Biological Effects" (O. Hutzinger, L. H. van Lelyveld, and B. C. J. Zoeteman, eds.), pp. 237–263. Pergamon, New York.

Zepp, R. G., and Schlotzhauer, P. F. (1979). Photoreactivity of selected aromatic hydrocarbons in water. *In* "Polynuclear Aromatic Hydrocarbons" (P. W. Jones and P. Leber, eds.), pp. 141–156. Ann Arbor Sci. Publ., Ann Arbor, Michigan.

2

Effects of Oil on Marine Mammal Populations: Model Simulations

Jerry M. Neff
Battelle Ocean Sciences
Duxbury, Massachusetts 02332

Introduction

The effect of an oil spill on a population of marine mammals can be assessed in a number of ways. One is to use a mathematical model that incorporates information on the behavior of spilled oil with data on the life history of the species in question. Also needed is some understanding of direct and indirect effects of oil on individuals—the kind of information usually generated from field observations and experimental studies. The advantage of a computer model is that it always produces a result: a prediction of impact. Modeling may be applied as part of risk assessment when developing offshore oil reserves.

But modeling has limitations: (1) simplified assumptions must be made to integrate processes that are highly variable and complex; (2) the prediction is only as reliable as the data used to drive the model; and (3) when models are linked (that is, the output from one is used as input to another), errors in the first will be propagated and perhaps magnified. Perhaps the greatest utility of a model is that it will identify gaps in data and will help to determine where research support should be directed.

Computer models have been developed and evaluated to predict the trajectory and fate of oil spilled in the marine environment (Samuels *et al.*, 1983); others have dealt with the population dynamics and behavior of marine mammals (Swartzman, 1984). There have been four attempts to

Sea Mammals and Oil: Confronting the Risks
Copyright © 1990 by Academic Press, Inc.
All rights of reproduction in any form reserved.

combine these two types of models:

1. "A Risk Analysis Model for Marine Mammals and Seabirds: A Southern California Bight Scenario" (Ford, 1985);
2. "Simulation Modeling of the Effects of Oil Spills on Population Dynamics of Northern Fur Seals" (Reed *et al.*, 1986);
3. "Computer Simulation of the Probability that Endangered Whales Will Interact with Oil Spills" (Reed *et al.*, 1987);
4. "A Simulation Model for Assessing the Risks of Oil Spills to the California Sea Otter Population and an Analysis of the Historical Growth of the Population" (Brody, 1988).

Marine Mammals in the Southern California Bight

Three models were linked in this study: the Oil Spill Risk Assessment Model (OSRAM), the Short Term Oil Response Model (STORM), and the Oil Spill Population Response Model (OSPREY) (Fig. 2–1) (Ford, 1985). OSRAM is an oil spill trajectory model used to estimate the probability that a spill will occur at a given location and contact a specified region (Lanfear *et al.*, 1979). In this analysis, OSRAM was used to generate trajectories and sizes of oil slicks originating from groups of existing leases and potential crude oil transport routes in the Southern California Bight. The purpose of STORM was to estimate the fraction of the population that would die as a result of a specified oil spill. OSPREY was used to assess the long-term effects of such mortality on population size and recovery rate.

The subjects of these modeling efforts were the northern fur seal, the northern elephant seal, the California sea lion, and the common dolphin. The study area was the Southern California Bight, the body of water bounded by the Southern California mainland to the east, the California Current to the west, Point Conception to the north, and the Mexican border to the south (Fig. 2–2).

Model Structure

Detailed descriptions of OSRAM are provided by Lanfear *et al.* (1979), Lanfear and Samuels (1981), and Smith *et al.* (1982). In brief, OSRAM uses wind and current data to simulate the trajectory of the center of an oil slick, given its source and the time of year. The model is stochastic, reflecting the extreme variability of typical wind patterns. Thus, no two simulated trajectories are identical. The standard version of OSRAM was modified by incorporating a submodel to simulate the radius of the slick and slick weathering (Mackay *et al.*, 1980). The volume of a spill was selected randomly from a log-normal distribution of spill volumes from oil produc-

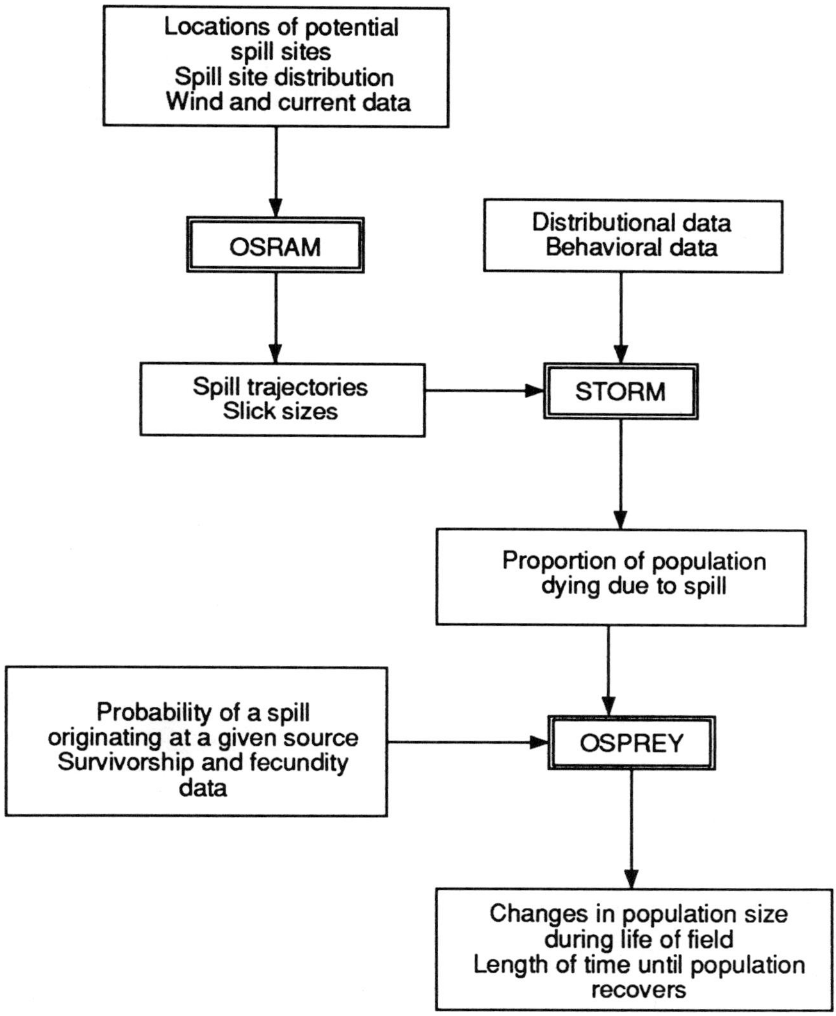

Figure 2-1
Flow diagram showing the relationship among the three models used in the analysis of risks to marine mammals from oil spills in the Southern California Bight.

tion platforms, pipelines, and oil tankers, based on historic spill volume statistics (Lanfear and Armstrong, 1983).

Spill trajectories were modeled for each of four seasons. Each spill was simulated from 1 of 14 production platform sites or from 1 of 14 segments of transportation (tanker or pipeline) routes in the study area

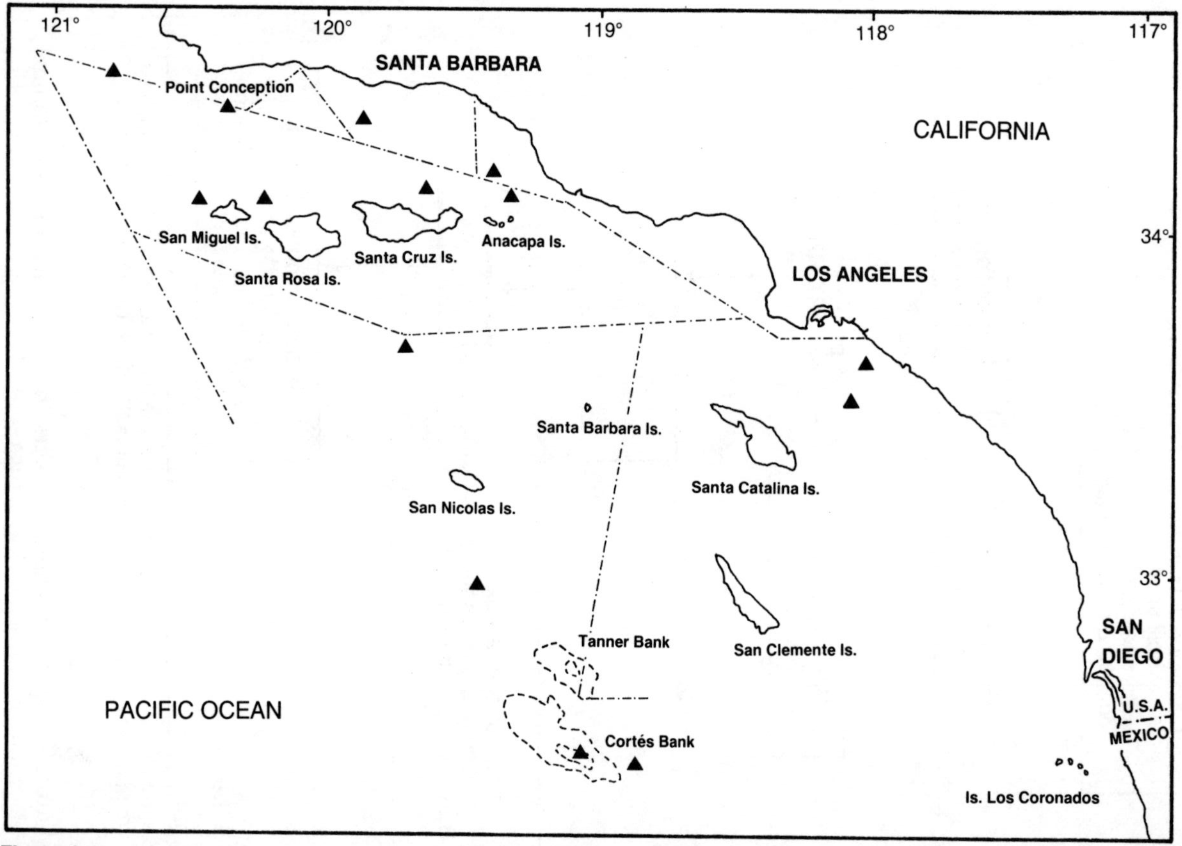

Figure 2-2

Map of the Southern California Bight, showing locations of principal features, platform sites (▲), and transport routes (-·-·-) used in the model.

(Fig. 2–2). Platform sites were considered to be points; transport segments were considered to be lines with equal probability of a spill occurring at any location along the line. Locations of platform sites, pipelines, and transport routes were based on groups of existing leased tracts on the continental shelf off southern and central California (LaBelle *et al.*, 1983a). For each of the 28 possible spill sites and each of the four seasons, 100 oil spill simulations were carried out. The output of each OSRAM simulation described the latitude and longitude of the center of the simulated oil slick and its radius at 3-hour time intervals. A simulation was terminated when (1) the center of the slick struck land, (2) the slick had weathered to 1% of its original volume, or (3) 30 days had elapsed from the time of the spill.

It was also necessary to estimate the probability of spills of different sizes occurring at a given location. These probabilities were based on the production at a given platform site, or the volume of oil transported along a particular tanker or pipeline route. Estimates of the volumes of oil produced, transported, and spilled were based on a preliminary version of the scenario of LaBelle *et al.* (1983a). It was assumed that the lifespan of a field was 25 years and that spill rates were the same as the historic rates of spills from platforms, tankers, and pipelines (Lanfear and Armstrong, 1983).

The output of OSRAM for a large number of spill trajectories from a particular location was then plotted as a contour map of index values representing the time after the spill that oil was present in a given region of the Southern California Bight. A region with a twofold index will be twice as likely to contain a slick from the specified source and, therefore, animals utilizing that area will be twice as likely to encounter the slick.

Using output from OSRAM, STORM estimated the number of animals that would contact the slick. Recognizing the variable consequences of such contact, for simplicity's sake it was assumed that if a marine mammal became oiled, it died. Direct contact between most species of pinnipeds (this volume, Chapter 4) and cetaceans (this volume, Chapter 6) and oil does not necessarily lead to mortality. Nevertheless, use of this worst-case scenario did not greatly affect the outcome of the modeling. Sensitivity analysis of the model revealed that the most important factors in determining the predicted portion of the population affected by the spill are the time of the year, the timing of the stages in the life cycle of the population being modeled, and the size and trajectory of the oil spill. In other words, the outcome of a model run depended primarily on whether the animals and oil happened to be in the same place at the same time.

The OSPREY model then predicted the long-term effect on the population. Estimates of age-specific annual fecundity and mortality were used to project population size and structure from year to year. Each simulation consisted of three phases. First, the current population was projected over the lifespan of the field, assuming only natural mortality; then the exercise was repeated to include mortality due to an oil spill. The third step was to determine the time required for an oil-exposed population to recover to the size that it would have attained under normal conditions by the end of the lifespan of the field.

Results

Northern Fur Seals

There are two northern fur seal rookeries on San Miguel Island in the Southern California Bight: Castle Rock and Adams Cove. Strips of nearshore water 500 m wide adjacent to these two rookeries were treated as targets for oil spills. Seals occupy the rookeries from May through December and give birth in June and July; females and pups remain in the rookeries for about 10 weeks (this volume, Chapter 3). From January to April, the seals are dispersed at sea in the Eastern California Bight. Thus, the impact of an oil spill will vary significantly with time of year. The greatest mortality resulted when spills in the Santa Maria Basin and western Santa Barbara Channel came ashore at the Castle Rock rookery in June. The probability of such an occurrence was estimated to be 0.3%.

The chance of a spill causing greater than 1% mortality ranged from 26.9% for adult males to 32.7% for pups. The probability of a spill causing greater than 15% mortality is 7.7% for females and immature animals and 5.8% for adult males. There is a 5.8% chance that a spill during the lifespan of the field will kill 40–50% of the young of the year.

To assess mortality for the whole Southern California Bight population of northern fur seals, three levels of probability were considered: (1) Conceivable but very unlikely. This was the maximum damage predicted by any model simulation. (2) Improbable, but sufficiently likely to be considered since this level of damage was predicted by the model 5% of the time (i.e., 95% worst case). (3) Likely to happen at least 50% of the time. As expected, there was a large difference between the worst case, when oil struck a rookery during pupping season, and the 50% worst case (Table 2–1).

The STORM model predicted that an oil spill would most likely reduce the population of the northern fur seals in the Southern California Bight by 2.2%, with a range of 1.0 to 3.7%. According to the OSPREY analysis, the population would recover within a year and, under the worst case, within 4 years.

Table 2-1
Highest Mortality (in %) Due to a Single Spill Under Conditions Defined by Three Levels of Probability[a]

Species	50% Worst case[b]	95% Worst case[c]	Worst case encountered[d]
Northern fur seal	0.9	18.9	23.0
Northern elephant seal	0.0	0.0	0.0
California sea lion	0.9	3.5	7.4

[a] From Ford (1985).
[b] Conditions predicted by the model at least 50% of the time.
[c] Predicted 5% of the time.
[d] Maximum mortality predicted by any model simulation.

Northern Elephant Seals

In the Southern California Bight, rookeries are occupied by northern elephant seals 95% of the time. Because different age and sex classes have specific seasonal haul-out patterns, the model divided the population into three categories: adult males, adult females and juveniles of both sexes, and pups. Rookeries on San Nicolas Island, the north shore coves of Santa Barbara Island, and Point Bennett, Adams Cove, and Tyler Bight on San Miguel Island were chosen as targets for a spill. Here, only pups were considered to be vulnerable because it was unlikely that an adult or juvenile encountering oil on shore would die.

The model predicted very little impact on elephant seals. There was no probability that a spill would kill more than 1% of the immature or adult animals, because they spend much time at sea and are relatively insensitive to oiling (this volume, Chapter 4). The probability that one or more spills would kill at least 1% of pups was 1.2%. Since the population is growing at a rate of approximately 13% per year, the estimated loss of pups would have no effect on the population as a whole (see Table 2–1).

California Sea Lions

At least 70,000 California sea lions occupy the inshore waters of the Southern California Bight for much of the year, most commonly in June and July when they give birth. Rookeries are located on San Miguel, San Nicolas, Santa Barbara, and San Clemente islands.

As for the northern elephant seal, the model divided the population into three age and sex classes: Adult females and immature animals, adult males, and pups. The probability of one or more spills resulting in greater than 1% mortality ranged from 11.7% for adult males to 35.6% for pups.

The scenario causing the greatest mortality of females and immature animals was a tanker or pipeline accident in the Santa Barbara Channel, spilling oil onto the rookeries on San Miguel Island during the breeding season in June. The resulting mortality was 7.4%. The worst case for adult males was 5.2% mortality following a similar spill in January that persisted for a month on Cortés Banks. A similar accident south of Santa Cruz Island in July, bringing oil to the San Nicolas Island rookery within 9 days, resulted in 14.2% mortality of pups.

The largest simulated spill-related mortality for the population as a whole ranged from 0.9% for the 50% worst-case situation to 7.4% for the worst case encountered (see Table 2–1). There was no seasonal tendency because the animals are uniformly distributed in the Bight throughout the year.

An oil spill would most likely reduce the population of sea lions by 1.2%, with a range of 0 to 4%. With an annual growth rate of 7.3% per year, the population would be expected to recover within 1 year, and at most within 3 years.

Common Dolphin

The distribution of common dolphins in the Southern California Bight is highly seasonal. From June through September, they range from the southern boundary to the Channel Islands; in fall, they are closer to San Nicolas and San Miguel islands; in the winter and early spring, they are restricted to the southern portion of the Bight.

The model was not used to predict mortality due to oil because there are no firm data linking the two for cetaceans (this volume, Chapter 6). Only the likelihood of contact was modeled. The risk was greatest in summer and lowest in winter. There was a 27.2% chance that one or more oil spills during the lifespan of the field would contact more than 1% of the common dolphins in the Southern California Bight. In the worst-case scenario, a spill from a platform between Santa Catalina Island and Long Beach remained in the area for most of October without coming ashore and covered an area of 52.8 km^2. In all, 24.6% of the dolphins contacted oil.

Northern Fur Seals in the Bering Sea

Two models were linked to predict the effects of a large oil spill on the Pribilof Islands herd of northern fur seals (Reed *et al.,* 1986) (Fig. 2–3). The first was a generic oil spill trajectory and weathering model. The

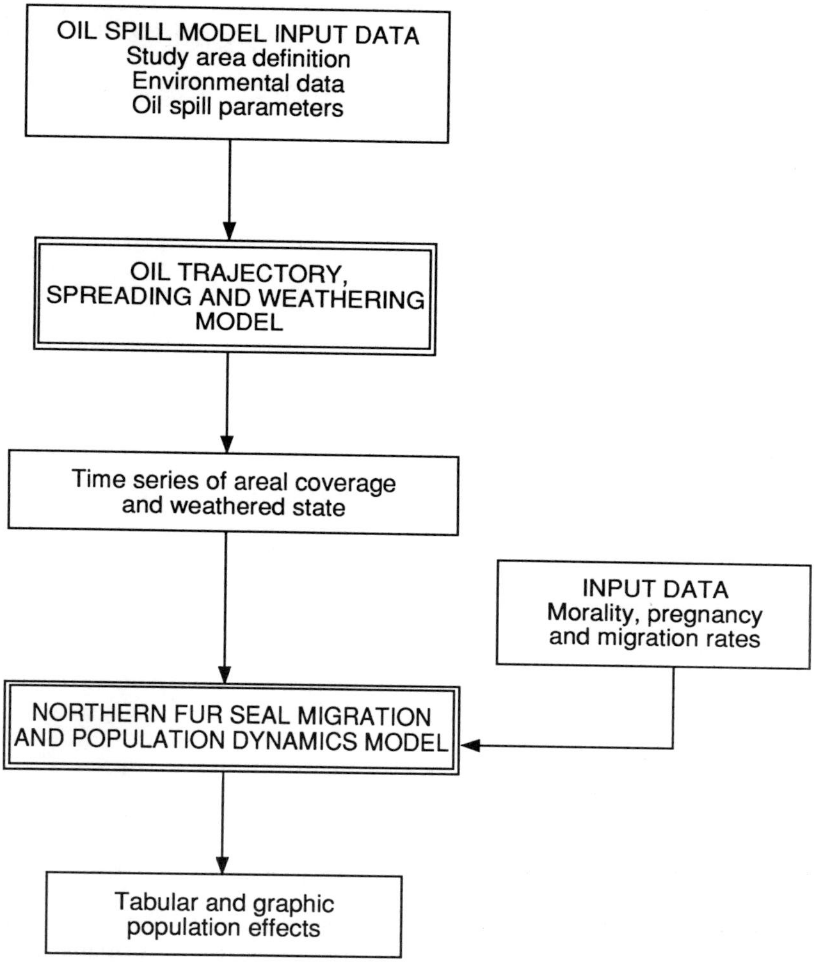

Figure 2-3
Schematic of linkages for fur seal–oil spill interaction model system.

second modeled population dynamics and movement patterns of fur seals in the Bering Sea.

Two hypothetical spills of 10,000 barrels each (420,000 gal) of Prudhoe Bay crude oil were simulated. The first was near Unimak Pass while seals were entering the Bering Sea from the Gulf of Alaska in the spring; the second was near the southern coast of Saint Paul Island in mid-July when the largest number of seals occupies rookeries (Fig. 2–4). These spill simulations were selected to maximize the likelihood of contact

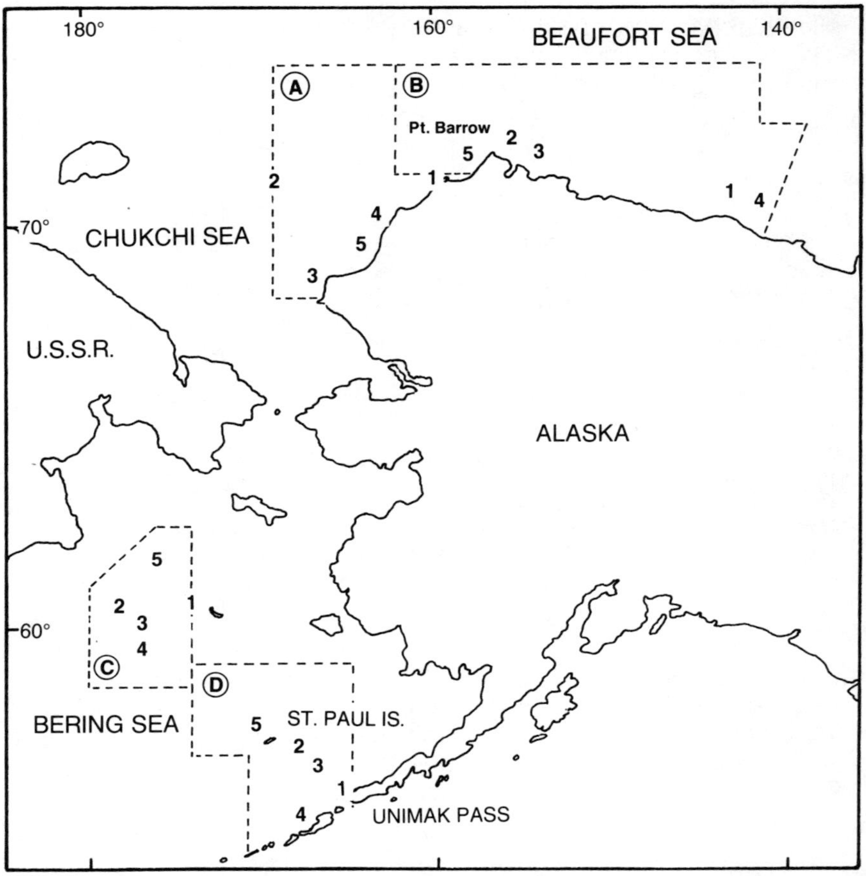

Figure 2-4
Oil spills were simulated in the Unimak Pass and near Saint Paul Island to determine the effect on northern fur seals. Five spill sites were selected in each of four areas considered for oil development to evaluate the risk of contact for bowhead and gray whales. A, Chukchi Sea; B, Beaufort Sea; C, Navarin Basin; D, St. George Basin.

between seals and oil, and should represent the worst-case scenarios. The probability of these spills occurring at the times and locations specified while fur seals are present is less than 2%.

Model Structure

The oil spill trajectory and weathering model was developed by Reed (1980) to simulate the interaction between spilled oil and fisheries species

(Spaulding *et al.*, 1982; Applied Science Associates, Inc., 1986). The spill model was run independently of the seal population dynamics model and assumed five spills, each of 2000 barrels (84,000 gal), over a 12-hour period. The event was followed for 10 days. Data on spill size and movements were then fed into the population dynamics model. Seal movements were tracked, and at fixed time intervals the position of the seals relative to that of the oil was determined.

The spill model accounted for evaporation of oil, spread of the slick, and dispersion into the water column (Mackay *et al.*, 1980; Payne *et al.*, 1984). Horizontal transport was computed by the hydrodynamic and wind/weather models of Applied Science Associates, Inc. (1986). The simulated July oil spill fouling Saint Paul Island was divided into two components: one coming ashore within 48 hours; the other after about 60 hours of weathering. The June spill near Unimak Pass remained at sea.

In the fur seal model, groups of animals were differentiated by sex, reproductive status, and age. The model assumed that if a pregnant or lactating seal died due to oiling, her pup died also. Because the fur seal population in the Bering Sea is not nutrient limited (seals are opportunistic feeders), possible effects of oil spills on food availability were not considered.

To run the model, mortality rates after oiling were arbitrarily set at 25, 50, 75, and 100%. Oiled seals that did not die were assumed to recover completely and to be no more or less sensitive to subsequent oilings. Thus, if a 50% mortality rate was assumed and the seal group encountered a spill twice, the net mortality of that group would have been 75% (half the survivors of a 50% kill). The model for the Unimak Pass spill assumed that migrating seals contacted oil only once; in the Saint Paul Island simulation, they contacted oil each time they entered or left the rookery.

Two population levels of fur seals were used in the simulations, one at equilibrium, comprising 1.16 million individuals, and the other at the 1986 estimate of 693,000 animals. The former is comparable to the estimated population size in 1979 (Lander, 1980, 1981). The 1986 estimate is lower, reflecting the loss of fur seals that become entangled and die in fragments of fishing nets (Swartzman, 1984; Fowler, 1985).

Results

The simulated spill in Unimak Pass fouled 0.05% of the males and 3.7% of the females in the population. Assuming a population of 1.16 million, the number of seals oiled in eight simulations averaged 29,364. Because the simulated spill coincided with peak migration of pregnant females to the rookeries, 94% of the seals oiled were females. Depending on the percent

Table 2-2
Numbers of Seals Oiled and Subsequently Dying as a Result of Oiling in Simulated Oil Spills Near Unimak Pass and Saint Paul Island Assuming Different Percent Mortalities for Oiled Seals[a]

Initial population	Mortality once oiled (%)	Unimak Pass		St. Paul			
				Oiled on rookeries		Not oiled on rookeries	
		Females	Males	Females	Males	Females	Males
Equilibrium	100	27,354	1,976	43,440	30,508	29,530	22,123
1.16 million	75	19,769	1,515	37,190	23,496	25,004	14,519
	50	13,245	1,010	29,090	18,579	18,114	10,254
	25	6,655	505	18,825	12,773	9,930	5,448
1986 Population	100	16,859	868	26,237	18,905	19,906	12,973
693,000	75	11,957	1,007	26,315	17,208	16,682	9,269
	50	7,974	682	22,097	13,564	12,258	6,777
	25	3,988	346	15,638	8,972	6,821	3,782

[a] From Reed *et al*. (1986).

mortality of oiled seals and the population estimate, predicted losses ranged from 4334 to 29,430 individuals (Table 2–2).

The Saint Paul Island spill simulation considered two alternatives: seals on adjacent rookeries were oiled, or they were not. The spill resulted in the oiling of a larger number of seals than the Unimak Pass spill, though the proportion of females was less (60%). In all, the spill fouled 4.0 or 6.2% of the females, and 5.3 or 6.4% of the males; the higher values assumed the oil reached the rookeries. In terms of animals, these values represent from 10,603–73,948 seals, depending on the initial size of the population and whether or not the rookery was oiled (Table 2–2). The model using the 1986 population level predicted that far fewer seals would be killed by oil than are currently estimated to die by net entanglement.

Recovery of the affected population to within 0.1% of the equilibrium population required 40–50 years, and about 20 years to within 1%. Only 5 years were required to overcome the smallest predicted impact which killed 12,000 seals. Most of the recovery occurred during the first few years.

Bowhead and Gray Whales in Alaskan Waters

Three models were linked to predict the probability that bowhead and gray whales would contact a spill in Alaskan waters (Fig. 2–5) (Reed and Jayko, 1986; Reed *et al.,* 1987). One generated a number of trajectories for spills at different times and conditions. Whale migrations were then simulated under the same conditions. The duration of contact was estimated from the number of surfacings and the average time at the surface, according to a "diving–surfacing" model. The final output of the model was the percentage of each population encountering oil and the probable number of interactions for each animal.

Model Structure

In this exercise, Reed and his co-workers used the same oil spill trajectory model for the whales as they had for the fur seals, but with a wider range of conditions. For each of five spill sites in the Navarin Basin, Saint George Basin, Chukchi Sea, and Beaufort Sea (see Fig. 2–4), 25 spill trajectories were generated for one or more seasons, varying the environmental conditions at the time of the spill, for a total of 500 simulations. For each scenario, the release date was selected randomly. Wind strength and

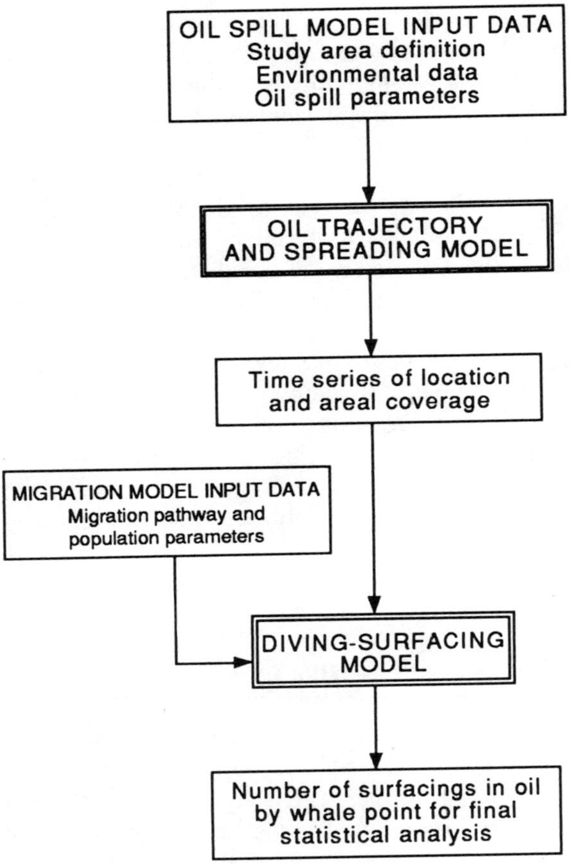

Figure 2-5
Schematic of linkages for the whale–oil spill interaction model system.

direction were established from historical data for that date, and ice conditions were set at light, average, or heavy. Half the simulations were run under average ice conditions, and 25% each for light and heavy conditions. Location of the ice edge was predicted from field observations (LaBelle *et al.,* 1983b).

Spills of 10,000 or 100,000 barrels (420,000 gal or 4,200,000 gal) were tracked for 10 days, thereby allowing time for weathering. When the model indicated that oil was trapped in or under ice, conditions slowing or preventing weathering, the spill was allowed to weather for a period that included an additional 10 ice-free days.

Whale migration models utilized available sighting data to define migration pathways, and speeds were calculated from field data on migration distance and time. The movement of a whale or group of whales was governed by a random walk algorithm that stochastically followed the migratory pathway. Whale migrations were begun on the first day of the month before the spill and were run until the end of the spill. The cumulative time each whale spent in oil was recorded periodically; there was no attempt to ascribe any effect resulting from an encounter.

Results

The model predicted that bowhead whales would contact oil in the Navarin Basin, Beaufort Sea, and Chukchi Sea (Table 2-3). In no instance did an encounter involve more than 1.9% of the population. The greatest number of contacts occurred in the Beaufort Sea, because bowheads spend much time each year feeding there (this volume, Chapter 5). According to the diving behavior model, most of the encounters involved fewer than 100 surfacings in oil-covered water.

Gray whales would contact oil in the Beaufort Sea and St. George Basin, and particularly in the Chukchi Sea where approximately 20% of the gray whale population feeds during summer. In St. George Basin, whales contacted oil while migrating to their feeding grounds in spring, and on their return in fall. Though virtually all the whales pass through Unimak Pass at these times, no more than 1.5% of the population contacted oil on each occasion. A few of these encounters lasted long enough for the whales to surface more than 300 times in oil.

Combining these results with estimates of the probability of oil spills in the Beaufort Sea from all sources (Minerals Management Service, 1985), there is a 51.8% chance that at least one bowhead whale will encounter oil during the 30- to 40-year lifespan of this oil field. The probability for the gray whale in the Bering Sea is 6.3%.

Sea Otters in California

A model was developed to assess the risk of oil spills to the California sea otter population (Brody, 1988). Beyond using existing data on the dynamics and demography of the population, the investigators undertook field studies to refine data on reproduction, growth rates, pup dependency, survival, age structure, movement patterns, and feeding behavior.

Table 2-3
Number of Simulated Spills Resulting in Encounters Between Oil and Whales and Percent of the Whole Population Encountering Oil within Ten Days of the Simulated Spills on the Alaskan Continental Shelf

		Number of encounters (out of 25 scenarios/site)		Percent of population encountering oil[a]	
Spill site	Season	Bowhead	Gray	Bowhead	Gray
Navarin Basin					
1	Feb 1–May 31	19	0	1.0	0
2	May 1–Nov 30	0	0	0	0
3	May 1–Nov 30	0	0	0	0
4	Feb 1–May 31	1	0	+	0
5	May 1–Oct 31	0	0	0	0
Beaufort Sea					
1	Aug 1–Oct 31	15	0	1.9	0
2	Apr 1–Jun 30	10	1	0.4	+
2	Aug 1–Oct 31	15	0	1.9	0
3	Aug 1–Oct 31	11	0	1.3	0
4	Aug 1–Oct 31	13	0	1.4	0
5	Apr 1–Jun 30	5	1	0.1	+
5	Aug 1–Oct 31	9	5	0.6	0.2
Chukchi Sea					
1	Apr 1–Jun 1	5	0	0.1	0
1	Jun 30–Oct 31	7	7	0.5	0.5
2	Aug 1–Oct 31	3	0	0.1	0
3	Mar 1–Jun 1	9	3	0.4	0.1
3	Jun 1–Oct 31	0	7	0	0.6
4	Jun 1–Oct 31	4	12	0.1	0.5
5	Jun 1–Oct 31	0	16	0	0.8
5	Oct 2–Jan 30	1	6	+	0.2
St. George Basin					
1	Mar 1–Jun 30	0	12	0	1.5
1	Aug 1–Dec 31	0	6	0	1.3
2	May 1–Oct 31	0	0	0	0
3	May 1–Oct 31	0	0	0	0
4	May 1–Oct 31	0	0	0	0
4	Nov 1–May 31	0	0	0	0
5	Apr 1–Nov 30	0	0	0	0

[a] +, A value greater than 0.0 but less than 0.1 percent.

Model Structure

The exercise used a stochastic simulation model of California sea otter population dynamics, with submodels giving information on distribution (OTDIST), movements (OTMOVE), and monthly (OTPOP) and yearly (LESLIE) projections of population status (Fig. 2–6). In operation, OTPOP created an initial population and monitored it for 3 to 4 years. At this point, OTDIST assigned each individual a position along the coast and passed this information to OTMOVE, which introduced a spill in Monterey Bay. OTMOVE moved animals along the coast during the spill, generating information on mortality. The model allowed the user to arbitrarily set values for mortality and avoidance. Results were fed to OTPOP, which monitored population recovery monthly for 2 years, and then on to LESLIE, which charted it annually for up to 50 years.

Results

Data were presented for a large spill introduced for 10 days along a 50-km section of the coast between Marina and Yankee Point in Monterey Bay. The initial population size in the area was set at its carrying capacity, 1600 otters. Mortality was assumed to be 100% of the otters remaining in the spill at the end of each day. No otters avoided the spill or emigrated from

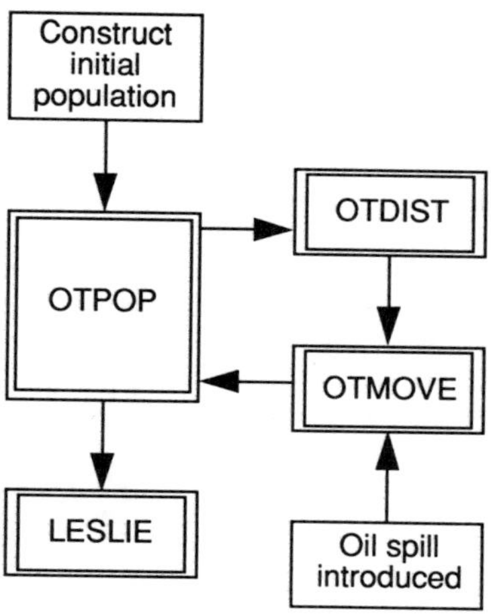

Figure 2-6
Schematic representation of the interrelation of the four submodels used to predict the potential effects of oil spills on California sea otter population dynamics.

it. Approximately 340 otters, or 18.6% of the population, died; nearly 70% were adult females and pups. Average time to recovery in 100 simulations of this event was 9.8 years. In 25% of the trials, the population approached but did not attain prespill size. Time to recover from a spill depended principally on the percentage of females killed, the status of the population in relation to its carrying capacity at the time, and the vicissitudes of the environment.

Summary

These modeling efforts are examples of the state of development of the approach. They can be quite sophisticated, incorporating a full range of environmental influences on oil and movements of animals that might contact it. When assumptions are made regarding mortality, the models can project the rate and degree to which a population recovers. A notable weakness in this scheme is that for most species, effects of contact are poorly understood, and the premise that contact is fatal is both extreme and results in gross overestimates of the impact of a spill. It is critical then to define more precisely how oil affects an animal. This can be done by combining direct observations and experimental studies with an understanding of life history, habitat use, social organization, demography, and other elements that can be used to measure the vulnerability of an animal to oil. The following chapters address these issues.

References

Applied Science Associates, Inc. (1986). "Ocean Circulation and Oil Spill Trajectory Simulations for Alaskan Waters: Spill Trajectory Simulations for Shumagin Oil and Gas Lease Sale No. 86," NOAA/OAD Contract No. WASV 85-00099. Natl. Oceanic Atmos. Admin., Washington, D.C.

Brody, A. (1988). A simulation model for assessing the risks of oil spills to the California sea otter population and an analysis of the historical growth of the population. *In* "Population Status of California Sea Otters" (D. B. Siniff and K. Ralls, eds), Rep. Contract No. 14-12-001-30033, pp. 191–274. U.S. Dept. of the Interior, Minerals Manage. Serv., Pacific Outer Continental Shelf Region, Los Angeles, California.

Ford, R. G. (1985). "A Risk Analysis Model for Marine Mammals and Seabirds: A Southern California Bight Scenario," Rep. OCS Study MMS 85-0104. U.S. Dept. of the Interior, 236 pp. Minerals Manage. Serv., Pacific Outer Continental Shelf Region, Los Angeles, California.

Fowler, C. W. (1985). An evaluation of the role of entanglement in the population dynamics of northern fur seals on the Pribilof Islands. *In* "The Fate and Impact of Marine Debris" (R. S. Shomura and H. O. Yoshida, eds.), Tech. Memo., Natl. Mar. Fish. Serv., NOAA-TM-NMFS-SWFC-54. Natl. Ocean. Atmos. Admin., Washington, D.C.

LaBelle, R. P., Lanfear, K. J., Banks, A. D., and Karpas, R. M. (1983a). An oil spill risk analysis for the Southern California Lease Offering (February 1984). Minerals Management Service Environmental Modeling Group. *Geol. Surv. Open-File Rep. (U.S.)* **83–563,** 1–115.

LaBelle, J. D., Wise, J. L., Voelkner, R. P., Schultze, R. H., and Wohl, G. M. (1983b). "Alaska Marine Ice Atlas." Arctic Environ. Inf. and Data Center, University of Alaska, Anchorage.

Lander, R. H. (1980). A life table and biomass estimate for Alaskan fur seals. *In* "Further Analysis of Pelagic Fur Seal Data Collected by the United States and Canada During 1958–1974. Parts 1 & 2" (H. Kajimura, R. H. Lander, M. A. Perez, and A. E. York, eds.), Rep. 23rd Annu. Meet. Standing Sci. Comm., North Pacific Fur Seal Commission, Natl. Oceanic Atmos. Admin./Natl. Mar. Fish. Serv., Seattle, Washington.

Lander, R. H. (1981). A life table and biomass estimate for Alaskan fur seals. *Fish. Res.* **1,** 55–70.

Lanfear, K. J., and Armstrong, D. E. (1983). A reexamination of occurrence rates for accidental oilspills on the U.S. Outer Continental Shelf. *In* "Proceedings of the 1983 Oil Spill Conference: Prevention, Behavior, Control, Cleanup," pp. 355–359. Am. Petrol. Inst., Washington, D.C.

Lanfear, K. J., and Samuels, W. B. (1981). Documentation and user's guide to the U.S. Geological Survey oilspill risk analysis model: Oilspill trajectories and the calculation of conditional probabilities. *Geol. Surv. Open-File Rep. (U.S.)* **81–316,** 1–95.

Lanfear, K. J., Smith, R. A., and Slack, J. R. (1979). An introduction to the oil spill risk analysis model. *Proc.—11th Annu. Offshore Technol. Conf.,* **11,** 2173–2175.

Mackay, D., Paterson, S., and Trudel, K. (1980). "A Mathematical Model of Oil Spill Behavior." Environ. Prot. Serv., Fish. Environ., Ottawa.

Minerals Management Service (1985). "Draft Environmental Impact Statement for the Proposed Five Year OCS Oil and Gas Lease Sale Schedule 1986–1991." U.S. Dept. of the Interior, Minerals Manage. Serv., Washington, D.C.

Payne, J. R., Kirstein, B. E., McNabb, G. D., Jr., Lambach, J. L., Redding, R. T., Jordan, R. E., Hom, W., de Oliveira, C., Smith, G. S., Baxter, D. M., and Goegel, R. (1984). Multivariate analysis of petroleum weathering in the marine environment—Subarctic. *In* "Environmental Assessment of the Alaskan Outer Continental Shelf—Final Reports of Principal Investigators," Vols. 21 and 22. U.S. Dept. of Commerce, Natl. Ocean. and Atmos. Admin., Outer Continental Shelf Environ. Assess. Program, Juneau, Alaska.

Reed, M. (1980). An oilspill fishery interaction model formulation and application. Ph.D. Thesis, University of Rhode Island, Dept. of Ocean Engineering, Kingston.

Reed, M., and Jayko, K. (1986). "Computer Simulation of the Probability that Endangered Whales Will Interact with Oil Spilled in the Navarin Basin Planning Area," Rep. Contract No. 14-12-0001-30076. U.S. Dept. of the Interior, Minerals Manage. Serv., Alaska Outer Continental Shelf Region, Anchorage.

Reed, M., French, D., Calambokidis, J., and Cubbage, J. (1986). "Simulation Modeling of the Effects of Oil Spills on Population Dynamics of Northern Fur Seals," Rep. OCS Study MMS 86-0045. U.S. Dept. of the Interior, Minerals Manage. Serv., Alaska Outer Continental Shelf Region, Anchorage.

Reed, M., Jayko, K., Bowles, A., Anderson, E., Leatherwood, S., and Spaulding, M. (1987). "Computer Simulation of the Probability that Endangered Whales Will Interact with Oil Spills," Rep. OCS Study MMS 86-0044. U.S. Dept. of the Interior, Minerals Manage. Serv., Alaska Outer Continental Shelf Region, Anchorage.

Samuels, W. B., LaBelle, R. B., and Amstutz, D. E. (1983). Applications of oilspill trajectory models to the Alaskan outer continental shelf. *Ocean Manage.* **8,** 233–250.

Smith, R. A., Slack, J. R., Wyant, T., and Lanfear, K. J. (1982). The oilspill risk analysis model of the U.S. Geological Survey. *Geol. Surv. Prof. Pap. (U.S.)* **1227,** 1–40.

Spaulding, M. L., Saila, S. B., Reed, M., Lorda, L., Walker, H., Isaji, T., Swanson, J., and Anderson, E. (1982). Oil spill fishery impact assessment model: Application to selected Georges Bank fish species. *Estuarine, Coastal Shelf Sci.* **16,** 511–541.

Swartzman, G. (1984). Present and future potential models for examining the effect of fisheries on marine mammal populations in the eastern Bering Sea. *In* ''Proceedings of the Workshop on Biological Interaction Among Marine Mammals and Commercial Fisheries in the Southeastern Bering Sea'' (B. R. Melteff and D. H. Rosenburg, eds.), Alaska Sea Grant Rep. 84–1, pp. 157–184. Alaska Sea Grant College Program, University of Alaska, Fairbanks.

3

Pinnipeds and Oil: Ecologic Perspectives

Ian A. McLaren
Biology Department
Dalhousie University
Halifax, Nova Scotia
B3H 4J1, Canada

Introduction

Pinnipeds share many characteristics with other marine mammals, and indeed with large mammals in general, especially in demographic features (Fowler and Smith, 1981), energetics (Lavigne *et al.*, 1986), and social behavior (Eisenberg and Kleiman, 1983). This allows us to draw on a wide range of empirical and theoretical literature to assess possible responses of pinnipeds to an environmental impact. Their amphibious nature poses special circumstances under which they face such threats.

The familiar fusiform bodies, with limbs modified as flippers, clearly reflect the pinniped's aquatic mode of life. They evolved from advanced terrestrial carnivores, related to the canid–ursid–mustelid line, perhaps from a common ancestral species. They are sufficiently coherent in most characteristics to be treated as a taxon. An excellent general account is by King (1983).

The three major kinds of pinnipeds are the hair seals (family Phocidae), the otariid seals (fur seals and sea lions; family Otariidae) and the walruses (family Odobenidae). Although intermediate in some respects, the walrus more closely resembles otariids in certain important ways. A prominent distinguishing feature of hair seals is their inability to rotate the hindflipper forward: theirs are fixed as "sculling" organs, while those of otariids and walruses can be turned forward in a more or less plantigrade position. The otariids and odobenids are accordingly more

Sea Mammals and Oil: Confronting the Risks
Copyright © 1990 by Academic Press, Inc.
All rights of reproduction in any form reserved.

Dense pelage of a male southern fur seal. Photo by D. W. Doidge.

mobile on rough terrain. Although hair seals can "slither" quite rapidly on sand, smooth rock, or especially ice, they must "hump" awkwardly over obstructions. When swimming, otariids use their large, propulsive foreflippers, placed close to midbody, while hair seals and walruses rely on their hindflippers, using the foreflippers for steering. Hair seals, walruses, and sea lions have short hair coats, especially sparse in the walrus. They are protected from excessive heat loss in part by a thick layer of blubber. Fur seals have a particularly dense underfur to trap air for insulation. Both blubber and thick fur are disadvantageous at high temperatures; with a few notable exceptions, pinnipeds are found in temperate to polar regions.

There is general agreement that there are 34 living species of pinnipeds. The 14 North American species are grouped in taxonomic categories in Table 3–1, which includes broad information on ranges and population status. Clearly, except for the Guadalupe fur seal, none is either very rare or excessively localized.

Distribution

Geographic distribution is a primary determinant of the probability of encounter with oil. In coastal and shelf waters of North America, pinnipeds occur from Mexico in the Pacific, up along the west coast through the Arctic Ocean and south to New England. At present only relatively small fractions of their ranges are leased or proposed for hydrocarbon exploration or production. Yet, because oil transport, even via the Arctic, is pervasive, I map the distributions of pinnipeds throughout North American waters (Figs. 3–1 to 3–18).

Life Histories

General Patterns

There is a common pattern to the annual cycles and habitat use of pinnipeds: they spend much of their lives at sea, but occupy land or ice to reproduce and often to molt (Table 3–2, Fig. 3–19). The timing and duration of haul-out behavior varies considerably. In species like the walrus, birth, mating, and molting are spread out in various segments of the population for more than half the year (Fig. 3–19). Some species almost never haul out, while others spend much time ashore between feeding forays at all times of year.

Land or ice is vital for the nurturing of young, except for pups of harbor seals (Lawson and Renouf, 1987), and perhaps the walrus (Fay, 1982) and the bearded seal (Burns, 1978) which are able to enter water soon after birth. Many species are selective in their choice of breeding ground, and for that reason, entire life histories feature seasonally synchronous reproductive activities at well-established sites. This may involve long-distance homing from feeding grounds to massive breeding colonies, much in the manner of seabirds, and with the same amplified risks of exposure to pollutants. Added to this is the likelihood that some individuals with strong site fidelity may refuse to abandon an area that has been contaminated.

Birth and Care of Young

There is a dichotomy between maternal behavior of phocids and that of otariids and walruses (Oftedal *et al.*, 1987). Attendance of young hair seals may be punctuated by brief departures of females, or their continuous presence during lactation periods that range from 4 days (Bowen *et al.*, 1985) to about a month. Walruses attend their young more or less continuously for much longer periods, while female sea lions and fur seals undertake lengthy feeding trips between suckling bouts (Gentry and Kooyman, 1986).

Table 3-1
Pinnipeds of North America with Estimates of Populations within Areas of Interest

Species	Breeding range	Population[a]	Status[b]	Sources[c]
Family Odobenidae				
Walrus	E. Canadian Arctic	25,000?	s?	Davis *et al.* (1980)
	Bering–Chukchi seas	160,000	s?	Estes and Gol'tsev (1984)
Family Otariidae				
Steller's sea lion	California	7,000	–	Mate and Gentry (1979)
	Oregon and Washington	3,000	s?	Everitt and Beach (1982)
	British Columbia	5,000(p)	s?	Obee (1984)
	Gulf of Alaska	103,000	s?	Loughlin *et al.* (1984)
	Aleutians, Bering Sea	93,000	–	Loughlin *et al.* (1984)
California sea lion	California	62,000+	+	Le Boeuf *et al.* (1983); Bonnell and Ford (1987)
	Mexico	83,000	+	Le Boeuf *et al.* (1983)
Guadalupe fur seal	Mexico	1,000+	+	Fleischer (1978)
Northern fur seal	Pribilof Islands	1,300,000	–	Lander (1981)
	San Miguel Island, California	7,000(p)	+	Cooper and Stewart (1982)
Family Phocidae				
Northern elephant seal	California, Mexico	60,000+	+	Le Boeuf (1982)
Bearded seal	E. Canadian Arctic	100,000?	s?	McLaren (1958b)
	W. Canadian Arctic	3,000+	s?	Stirling *et al.* (1977)
	Bering–Chukchi seas	300,000	s?	Burns (1981a)
Hooded seal	E. Canada, Davis Strait	366,000(p)	+	Bowen *et al.* (1987b)
Ringed seal	E. Canadian Arctic	1,000,000?	s?	McLaren (1958c)
	Beaufort–Chukchi seas	40,000+	s?	Frost and Lowry (1984); Stirling *et al.* (1977)
	Bering Sea	1,250,000?	s?	Lowry and Frost (1981)

Spotted seal	Bering–Chukchi seas	225,000?	s?	Lowry and Frost (1981)
Harbor seal	New England	10,000+[d]	+	Payne and Schneider (1984)
	E. Canada	13,000+	+?	Boulva and McLaren (1979)
	Labrador	?	?	Mansfield and Sergeant (1960)
	E. Canadian Arctic	100s?	−?	Mansfield (1967b)
	S. Alaska	67,000+	s?	Everitt and Braham (1980); Calkins and Pitcher (1983)
	Aleutians	?	?	Burns and Gol'tsev (1984)
	British Columbia	35,000?	s?	Bigg (1969b)
	Washington	8,000+	+?	Calambokidis *et al.* (1979)
	Oregon	3,000+	s?	Everitt and Beach (1982)
	California	6,000+	s?	Stewart (1980)
Harp seal	E. Canada	2,250,000(p)	+	Roff and Bowen (1983, 1986)
Ribbon seal	Bering–Chukchi seas	100,000?	s?	Burns (1971)
Gray seal	E. Canada	70,000	+	Zwanenburg *et al.* (1985)

[a] A lack of symbol indicates that the estimate is a mean based on actual counts with attempts at statistical analysis; + indicates that such mean estimates were believed by the source to be minimal; ? after a number indicates that the estimate is considered by the source to be highly approximate, or is not based on stated sampling procedures or statistical analyses; a ? without a number indicates that no estimates have been found, although some information on the population is given in the source publications; (p) indicates that the estimate is based on counts of young, here multiplied by 4.5 as estimates of total populations.

[b] The symbol s indicates that the source publications imply that the population is thought to be more or less stationary; + indicates that the population is thought to be increasing, and − that it is decreasing; ? indicates that uncertainty is expressed about status.

[c] The source populations are generally the latest original references to population size and status. Secondary sources are used where original estimates are qualified or where the primary sources are relatively inaccessible.

[d] Probably largely nonbreeding migrants from Atlantic Canada.

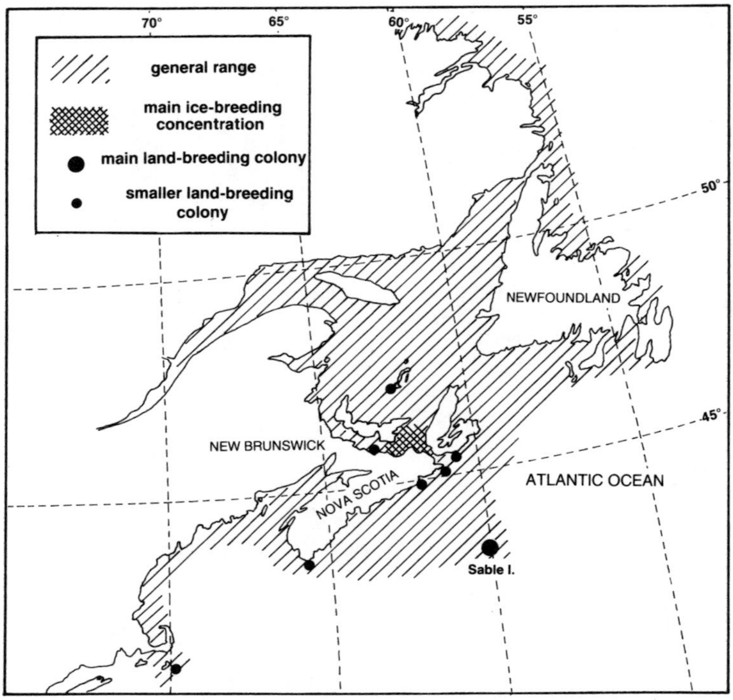

Figure 3-1
Distribution of the gray seal in the western North Atlantic (after Mansfield and Beck, 1977).

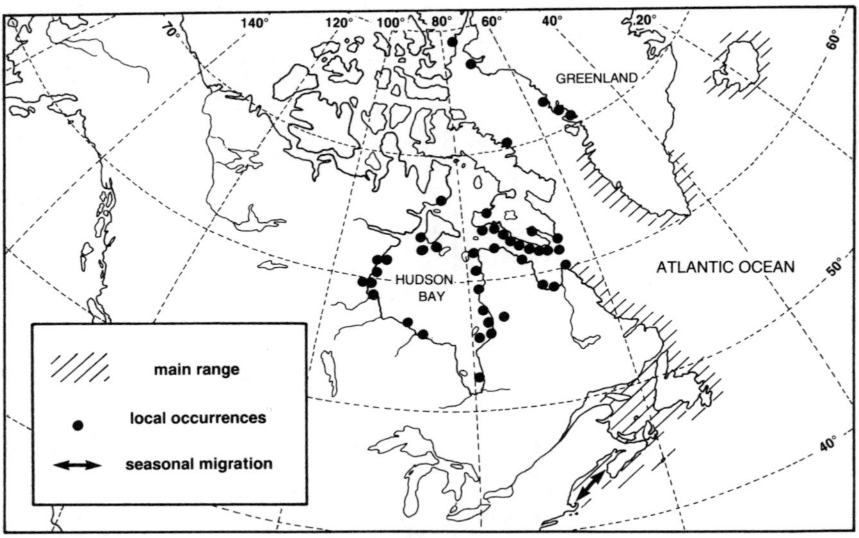

Figure 3-2
Distribution of the harbor seal in the western North Atlantic (after Mansfield, 1967a,b).

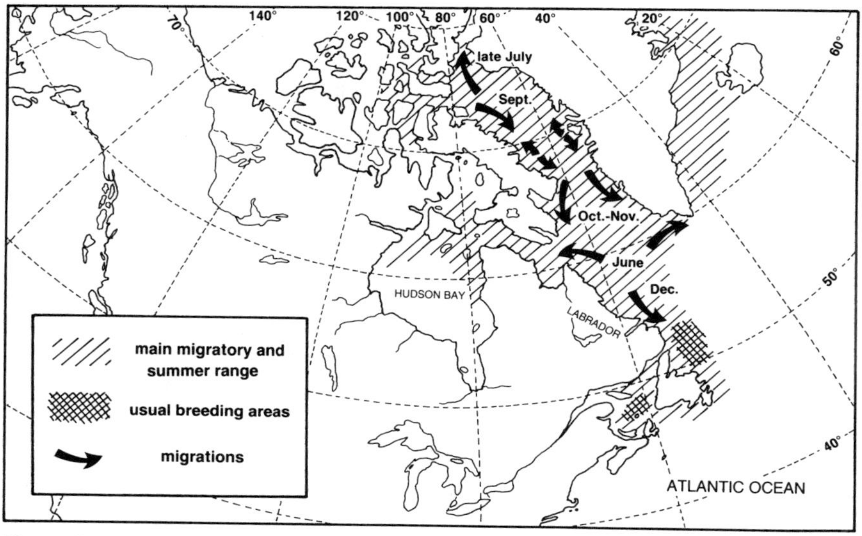

Figure 3-3
Distribution of the harp seal (after Sergeant, 1965; Davis *et al.*, 1980).

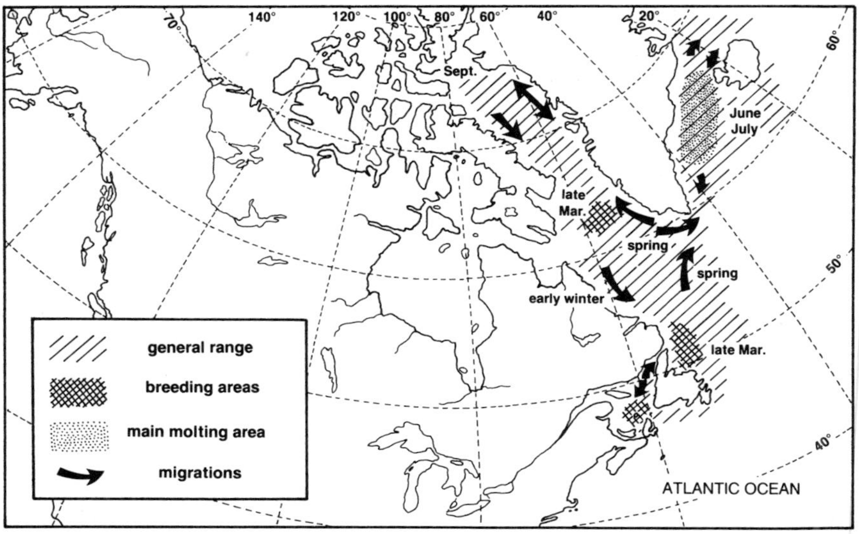

Figure 3-4
Distribution of the hooded seal. Many extralimital records, as far as Alaska and Florida, have been omitted (after Davis *et al.*, 1980; Reeves and Ling, 1981).

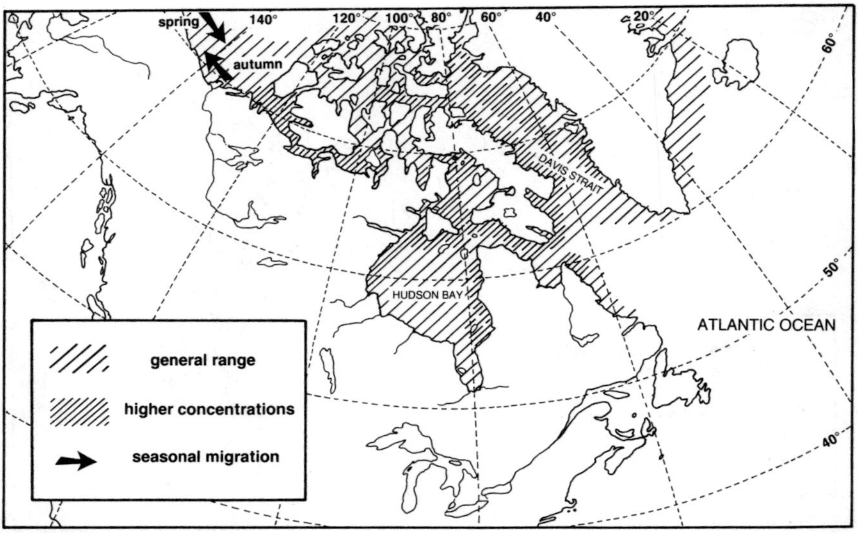

Figure 3-5
Distribution of the ringed seal in northern Canada (after McLaren, 1958c; Smith, 1975; Davis *et al.*, 1980; Finley *et al.*, 1983).

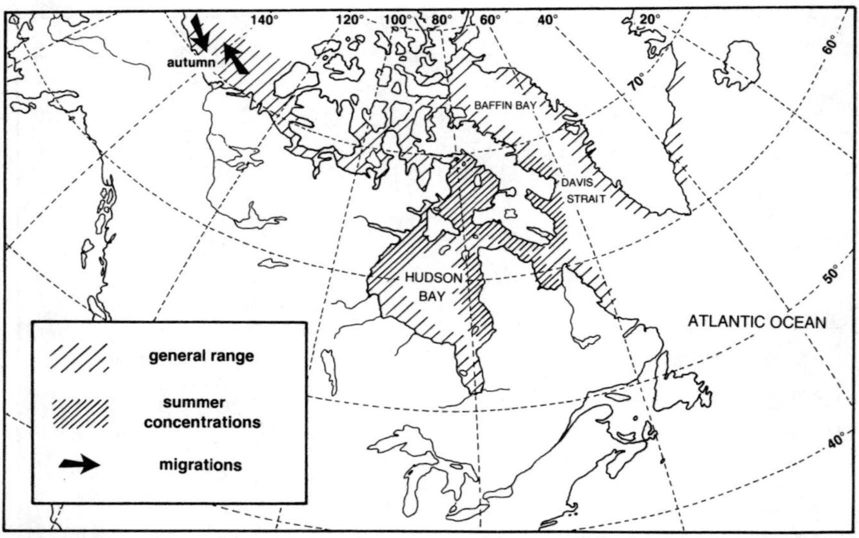

Figure 3-6
Distribution of the bearded seal in the Canadian Arctic (after Mansfield 1967a; Davis *et al.*, 1980).

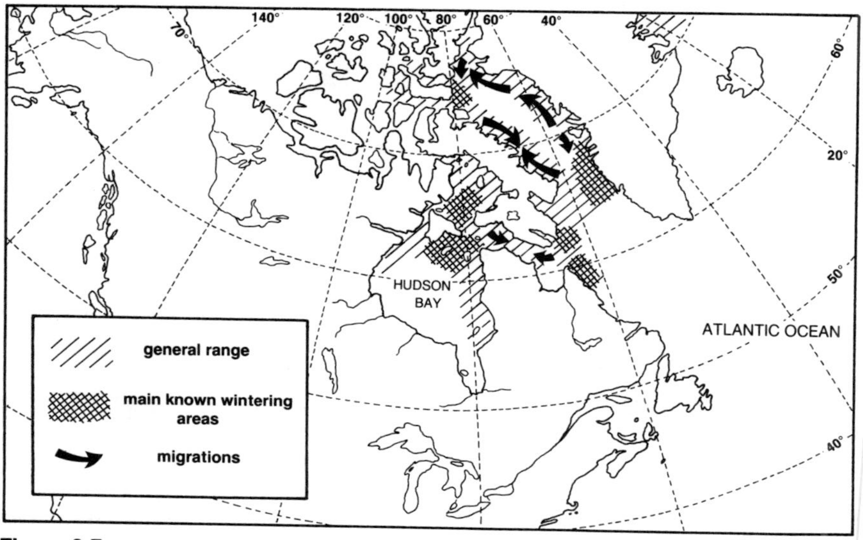

Figure 3-7
Distribution of the Atlantic walrus (after Davis *et al., 1980).*

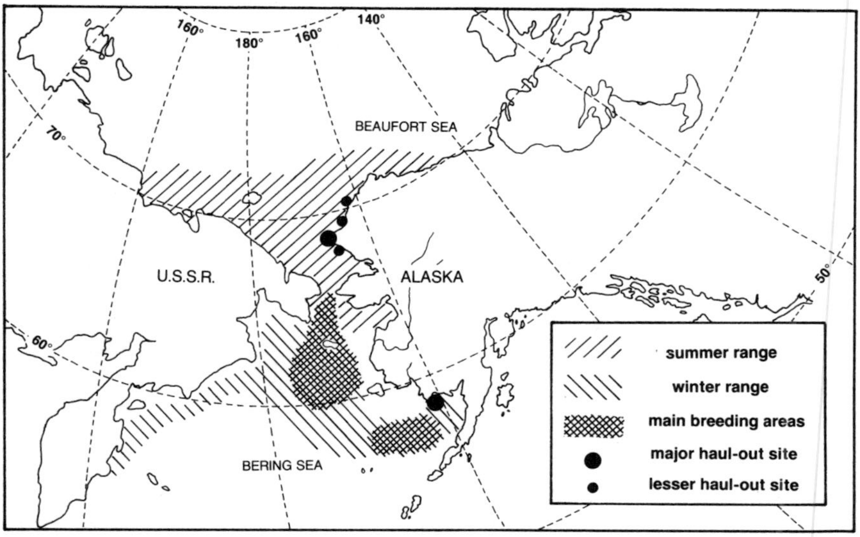

Figure 3-8
Distribution of the Pacific walrus in the Bering and Chukchi Seas (after Fay, 1982; Frost *et al.,* 1983).

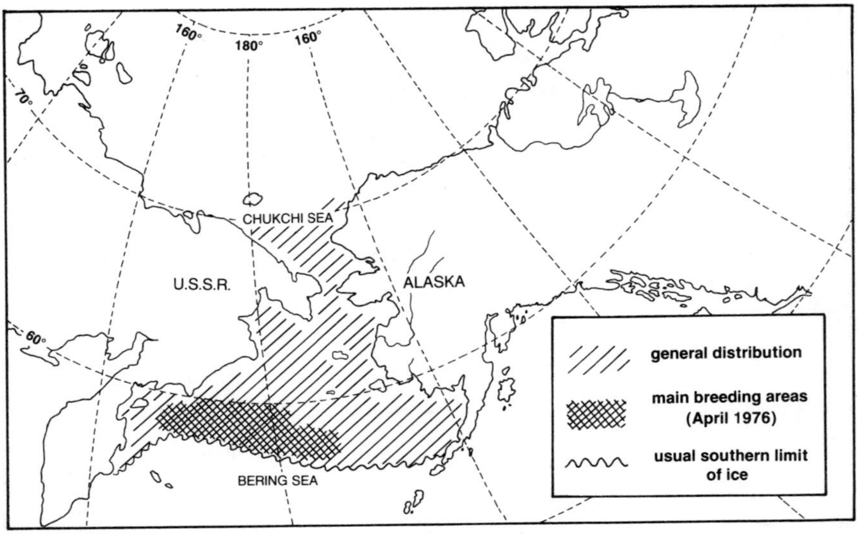

Figure 3-9
Distribution of the ribbon seal (after Burns, 1981a; Braham *et al.*, 1982; Frost *et al.*, 1983).

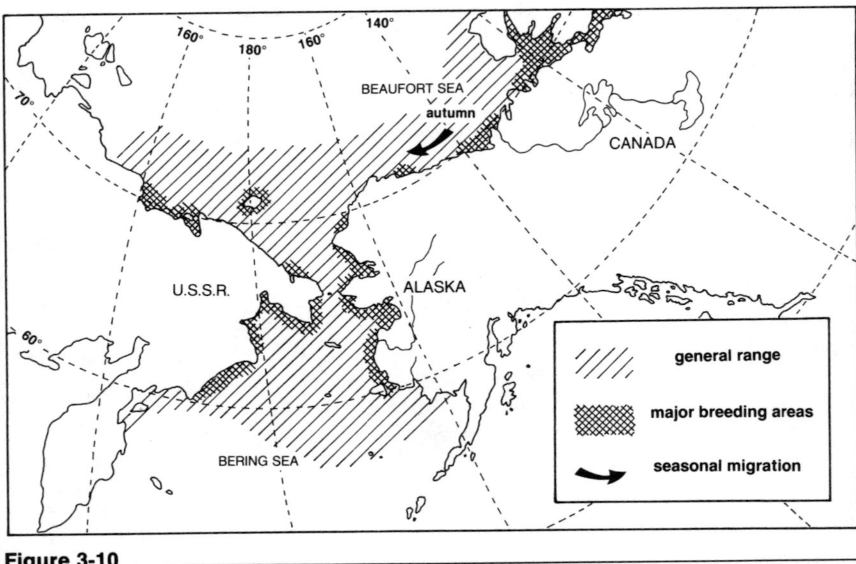

Figure 3-10
Distribution of the ringed seal in the Bering, Beaufort, and Chukchi Seas (after Burns, 1978).

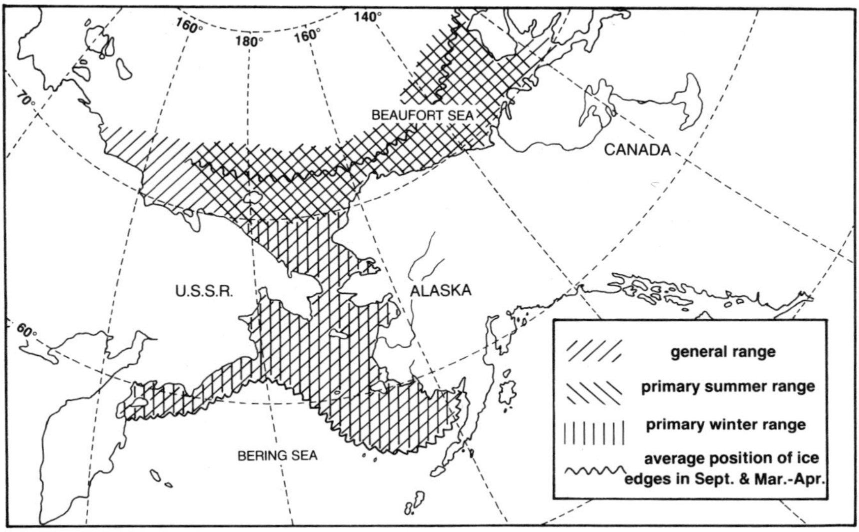

Figure 3-11
Distribution of the bearded seal in the Bering, Chukchi, and Beaufort Seas (after Burns, 1978; Braham *et al.*, 1982).

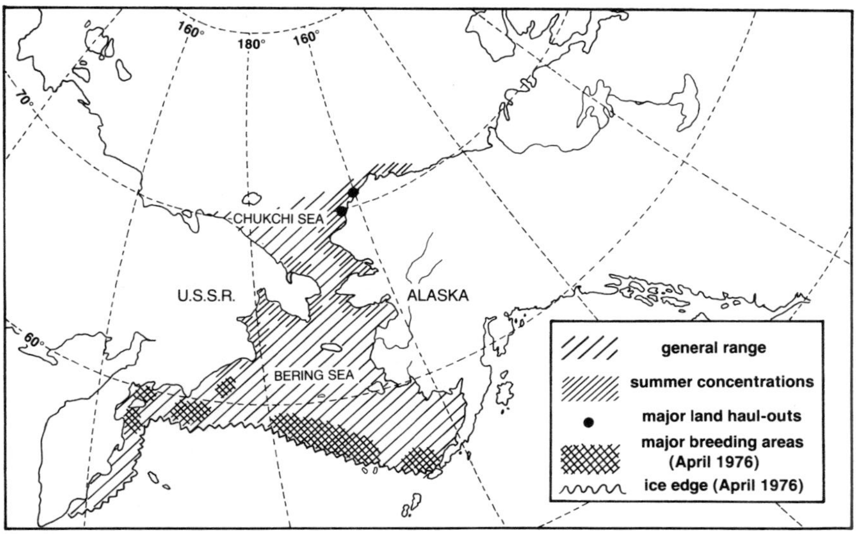

Figure 3-12
Distribution of the spotted seal (after Bigg, 1981; Braham *et al.*, 1982; Davis and Thomson, 1984).

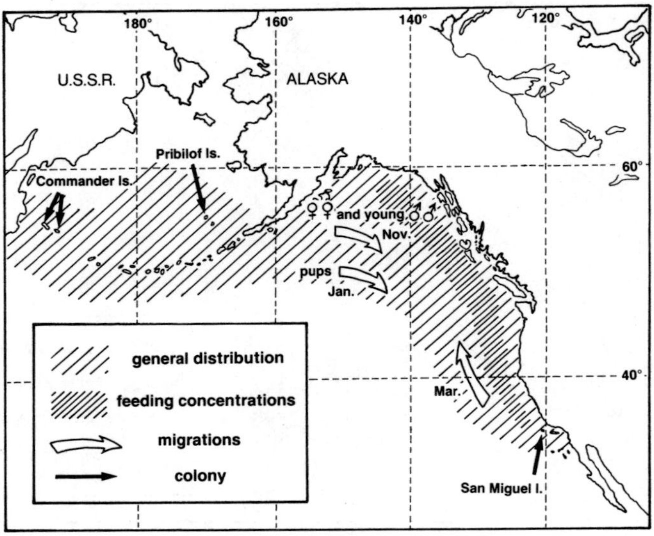

Figure 3-13

Distribution of the northern fur seal (after Fiscus, 1978; Braham *et al.*, 1982).

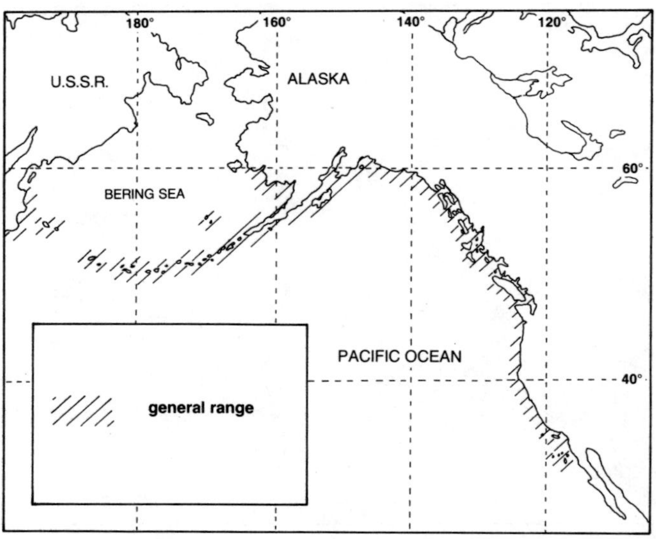

Figure 3-14

Distribution of the harbor seal in the Pacific (after Bigg, 1981; Burns and Gol'tsev, 1984).

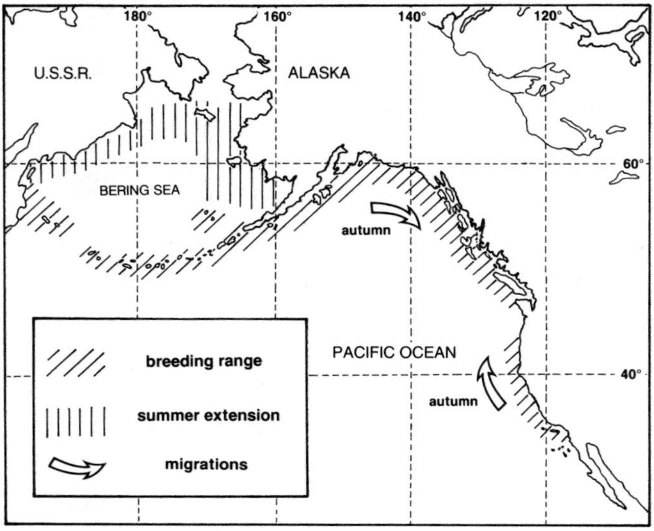

Figure 3-15
Distribution of the Steller's sea lion (after Schusterman, 1981).

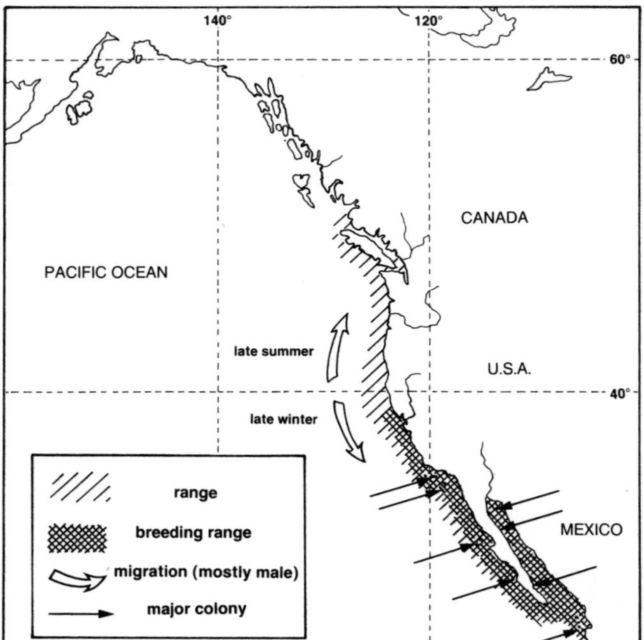

Figure 3-16
Distribution of the California sea lion (after Odell, 1981; DeMaster *et al.*, 1982; Le Boeuf *et al.*, 1983).

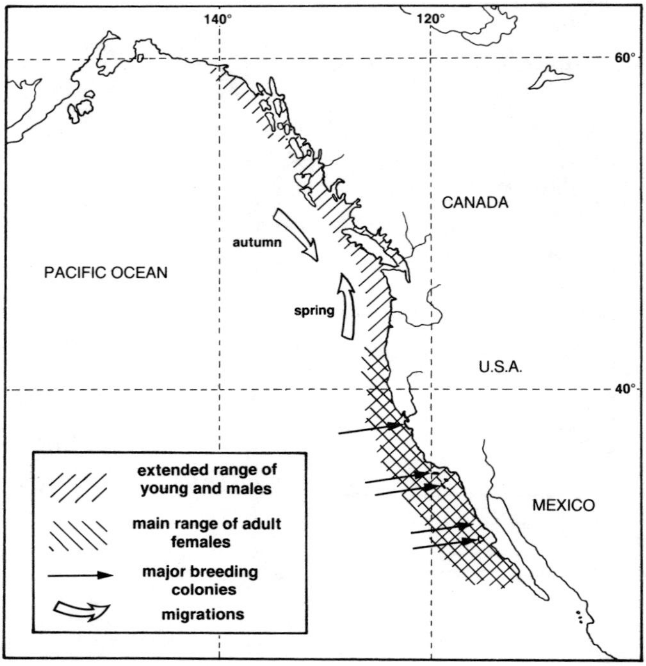

Figure 3-17
Distribution of the northern elephant seal (after DeLong, 1978).

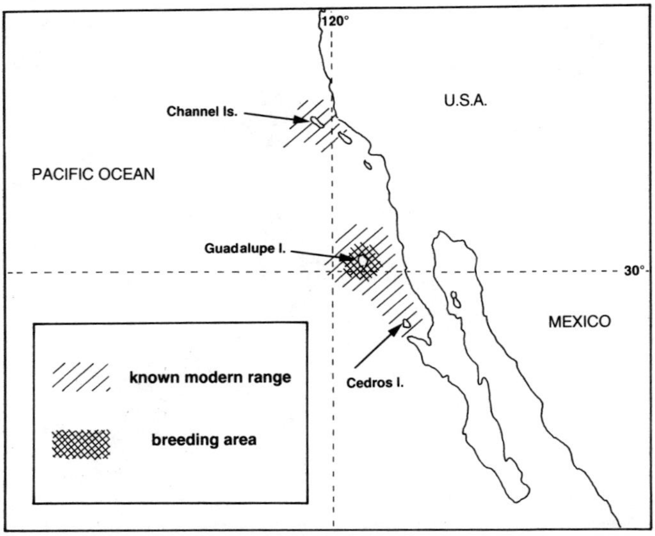

Figure 3-18
Distribution of the Guadalupe fur seal (after Fleischer, 1978).

Table 3-2
Broad Patterns of Habitat Use by North American Pinnipeds[a]

Species	Breeding haulouts	Molting haulouts	Haulouts at other times	Aquatic habitats–ranges[b]
Walrus	Pack ice	Pack ice or land	Pack ice or land	Coastal ↔ pelagic (shelf)
Steller's sea lion	Land	Not needed?	Land	Coastal ↔ pelagic
California sea lion	Land	Not needed?	Land	Coastal
Northern fur seal	Land	Not needed?	Land, rarely	Coastal → pelagic
Guadalupe fur seal	Land	Not needed?	Land	Coastal
Northern elephant seal	Land	Land	Land, uncommonly	Coastal ↔ pelagic
Bearded seal	Pack ice	Pack ice	Pack ice, if available	Coastal ↔ pelagic (shelf)
Hooded seal	Pack ice	Pack ice	Pack ice, if available	Coastal ↔ pelagic
Ringed seal	Fast and pack ice	Fast and pack ice	Pack ice, if available	Coastal
Spotted seal	Pack ice	Pack ice	Pack ice and land	Pelagic → coastal
Harbor seal	Land	Land	Land	Coastal
Harp seal	Pack ice	Pack ice	Pack ice, if available	Coastal ↔ pelagic
Ribbon seal	Pack ice	Pack ice	Pack ice, if available	Pelagic
Gray seal	Pack ice, land	Land	Land	Coastal ↔ pelagic

[a] From various routine accounts.
[b] Migratory movements after reproduction indicated by directional arrows. Localized or short-term exchanges by double-ended arrows.

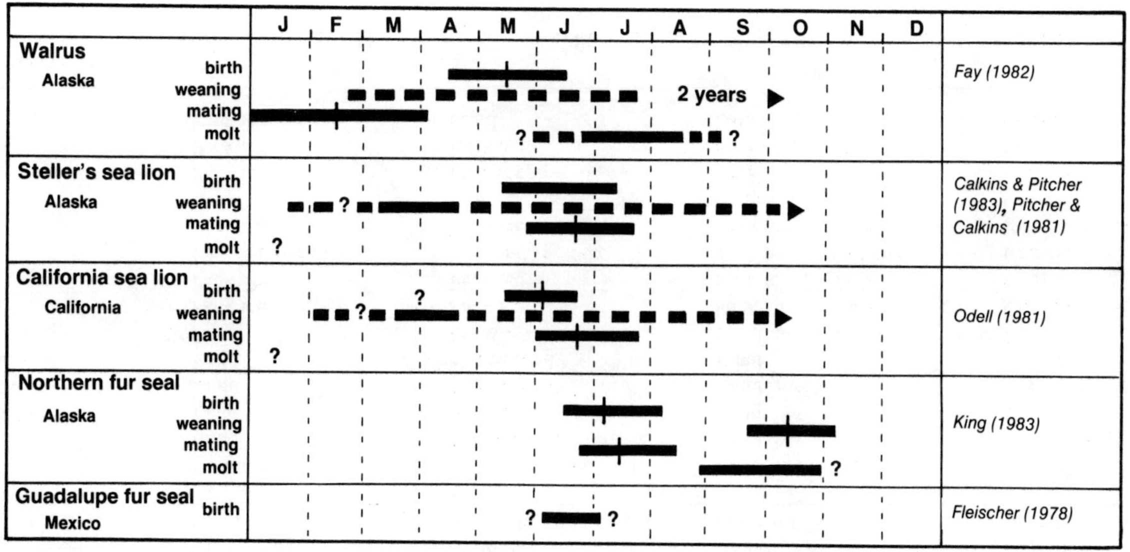

Figure 3-19

Life history tables for pinnipeds occurring in North American waters. Only main references given. (*Figure continues*)

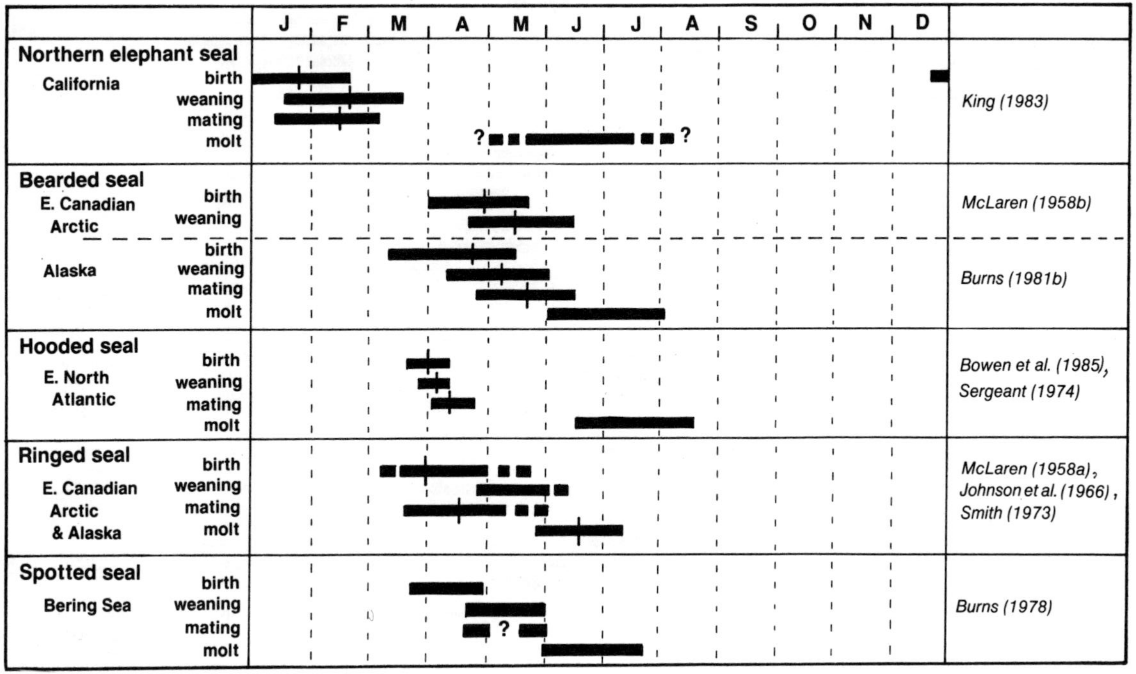

Figure 3-19 (*Continued*)

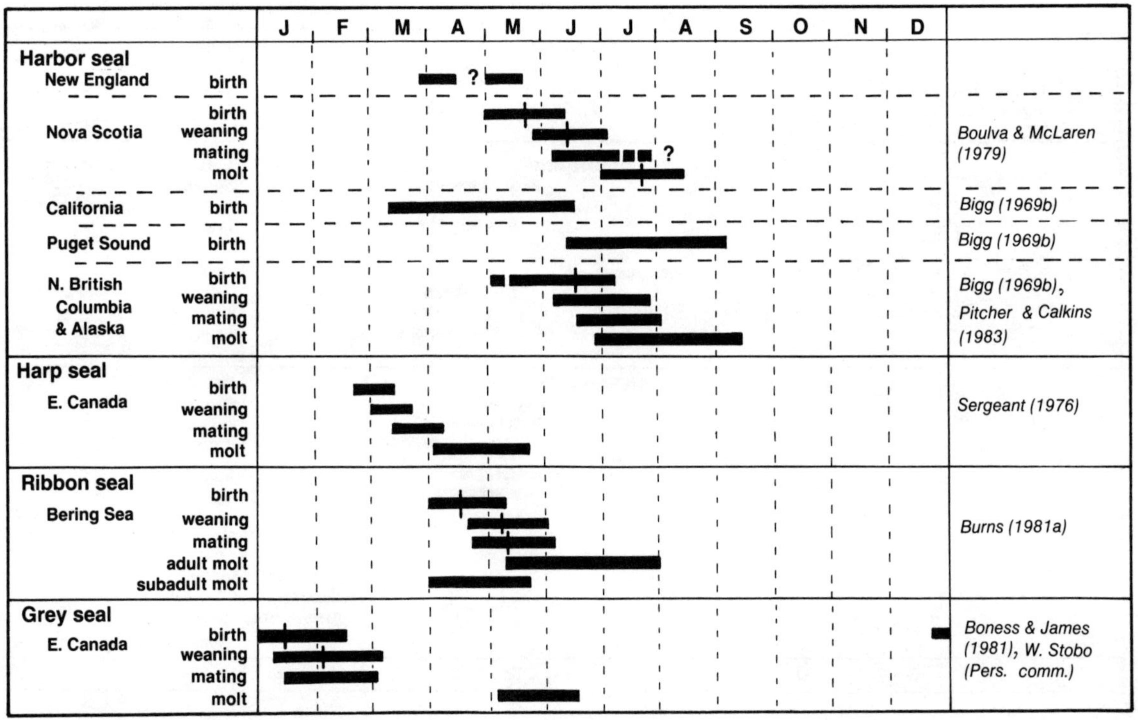

Figure 3-19 (*Continued*)

Sea lions resting on a haul-out ledge in Baja California. Photo by B. Würsig.

Growth and Maturation

Growth rates and body sizes of pinnipeds (Table 3–3) are presumably adapted to environmental circumstances, but no obvious biogeographic rules are evident. For example, the huge walrus and the much smaller ringed seal occur together in polar waters; the even larger elephant seal is found in subtropical Mexico; and the small harbor seal ranges without obvious differences in body size from the high Arctic to Baja California. As a rule, females double their length, and therefore increase their core weight about eightfold between birth and full size (see Table 3–3). The relative weights of fully grown animals, often inaccurately recorded in the literature, can also be approximated from the cubes of lengths given in Table 3–3. Rate of growth and final body size of some males are probably driven by the advantages of large size for threat and combat during breeding. This sexual dimorphism is pronounced among otariids and elephant seals in which males may weigh 2.5–5 times as much as females, and less so in the gray seal, in which full grown bulls weigh about 1.5 times more than females. Males of highly polygynous species may be potent when quite young (Spotte and Schneider, 1982), but have a sharp growth spurt at puberty and mate effectively only after reaching much greater body size. Males of "monogamous" species are generally the same size or smaller than females, even those which defend underwater or under-ice breeding territories. The male hooded seal seems anomalous, weighing some 1.6 times as much as a female. Although it defends individual females on pack ice, it is not strictly monogamous; aggressive males may be capable of mating with several females sequentially (Boness *et al.,* 1988).

Table 3-3
Sizes of North American Pinnipeds[a]

Species	Region	Newborn Length	Asymptotic length		Sources
			Male	Female	
Walrus	Alaska	115	311	260	Fay (1982)
Steller's sea lion	Alaska	106	293	230	Fiscus (1961); Calkins and Pitcher (1983)
California sea lion	California, Mexico	73	ca.225	ca.180	Gilmartin *et al.* (1976); Lluch (1969)
Northern fur seal	E. North Pacific	63	190	129	McLaren and Smith (1985)
Northern elephant seal	California	ca.150	ca.450	322	Le Boeuf (1979)
Bearded seal	Bering–Chukchi seas	131	223	223	Burns and Frost (1983)
Hooded seal	Greenland–E. Canada	93	225	200	Wiig (1985); Beloborodov and Potelev (1966)
Ringed seal	Bering Sea	63	143	132	Fedoseev (1965)
	Chukchi Sea	61	123	119	Fedoseev (1965); Johnson *et al.* (1966)
	Beaufort Sea	71	127	133	Smith (1987)
	S. Baffin Island	68	127	127	McLaren (1958a)
Spotted seal	Bering–Okhotsk seas	81	153	149	Tikhomirov (1968)
Harbor seal	E. Canada	78	157	149	Boulva and McLaren (1979)
	British Columbia	82	176	156	Bigg (1969a)
	S. Alaska	82	162	150	Pitcher (1977); Pitcher and Calkins (1983)
Harp seal	E. Canada	85	169	169	Stewart and Lavigne (1980); Innes *et al.* (1981)
Ribbon seal	Bering–Okhotsk seas	81	156	156	Shustov and Yablokov (1967); Fedoseev (1973)
Gray seal	E. Canada	108	202	238	Mansfield (1978)

[a] Lengths (in cm) measured in a variety of ways in some original sources have been converted to standard lengths and asymptotic (not maximum) lengths determined by fitted growth curves using methods given by McLaren and Smith (1985). Those qualified by ca. are unreliable, usually largest rather than asymptotic lengths.

Size differences among individuals and species could influence their response to environmental contamination. Large individuals with favorable surface to volume ratios might be more resilient. However, when large size is driven by sexual selection to the detriment of other components of fitness, some advantage might be lost. This is perhaps reflected in the reduced lifespans of large, polygynous species.

Diet and Feeding Tactics

Pinnipeds generally have broad opportunistic diets. The majority are piscivorous, but many also feed on cephalopods, planktonic crustaceans, and epibenthic organisms. Only the walrus and to a lesser extent the bearded seal feed primarily on burrowing bottom animals. Some North American pinnipeds consume seals or birds from time to time, though none is as voracious a carnivore as the Antarctic leopard seal. Lowry and Fay (1984) also document the remains of seals, mostly pups, in 8 of 645 stomachs of walruses from the Bering Strait and western Chukchi Sea. Steller's sea lions sometimes prey on harbor seals (Pitcher and Fay, 1982) and, perhaps regularly, on northern fur seal pups (Gentry and Johnson, 1981). The literature on apparently casual predation on seabirds by pinnipeds is summarized by Lucas and McLaren (1988).

In conclusion, significant amounts of hydrocarbons would probably not be consumed by pinnipeds in their food, since none of the prey is likely to accumulate residues. Exceptions are bearded seals and walruses foraging in heavily contaminated benthos and individuals of a number of species that might specialize in eating seals or birds (Lucas and McLaren, 1988), thereby consuming raw oil entrained in pelage or plumage. This probability would be enhanced if contamination rendered the prey more vulnerable to capture.

Habitat Use

Major Patterns of Distribution

The marine ranges of pinnipeds during the nonbreeding season fall into four somewhat overlapping categories: coastal versus pelagic, and ice-using (pagophilic) versus ice-shunning (pagophobic). Each category presents specific circumstances under which a species would encounter oil.

Some coastal species migrate regularly, others disperse during the nonbreeding season, while still others remain in the vicinity of breeding sites. Most species that stay inshore through the year are generalized, even opportunistic, predators, and their distributions are probably more influenced by availability of suitable hauling-out sites than by food requirements. Boulva and McLaren (1979) found a strong correlation on a local scale between abundance of harbor seals in Nova Scotia and the number of islets along the coastline. On a regional scale the abundance and species richness of pinnipeds around such places as the southern California islands have much to do with their suitability for safe hauling out. Of course coastlines with many islands, islets, and bars could also pose greater risks of oil spills through marine accidents. Furthermore, the enclosed topographies of bays, estuaries, and passages used by coastal pinnipeds can concentrate spilled oil.

Among coexisting coastal species, there are some indications of niche differences. Thus, on the West Coast, the two major coastal otariids are generalized, nocturnal feeders, but the Steller's sea lion ranges further offshore than the California sea lion (Fiscus and Baines, 1966), and the latter is seen increasingly in estuaries (Bayer, 1981). The harbor seal is also a generalist in diet, seldom ranges seaward, and is not notably nocturnal. Elephant seals evidently feed more frequently in deeper water on larger and more bottom-dwelling fishes than do the other coexisting pinnipeds (McGinnis and Schusterman, 1981). On the East Coast, the gray and harbor seals are piscivores, but the former ranges over the Scotian shelf and central Gulf of St. Lawrence, while the latter is rarely seen far out at sea. The general trend among these species is that the larger ones range farther offshore.

Pelagic species by definition spend the nonbreeding seasons in offshore waters where hauling out is possible only on ice. The distribution and movements of such species are more likely to be controlled by availability of food, in turn influenced by hydrography and submarine topography. For example, the northern fur seal is concentrated in winter along the outer continental shelf from British Columbia to California, where it feeds on schooling fishes and squids (Kajimura, 1984). The ribbon seal evidently summers largely near the edge of the Bering Sea shelf (Burns, 1981a). These distributions may be related to high productivity and are generally remote from oil exploration activity or concentrated oil-shipment routes. The distributions of the two benthic feeders, walrus and bearded seal, are constrained within the continental shelves, but can range quite far offshore, especially in the Bering Sea.

Species that produce young on ice are accordingly limited in distribution, in spite of occasional anomalies such as the parturition of a hooded seal on land (Richardson, 1975). Among North American species (Tables 3–2 and 3–4), only the ringed seal breeds on fast ice or pack ice (Finley *et al.,* 1983). Postreproductive hauling out by ice-using species is also variable (Tables 3–2 and 3–4). Some, like the hooded, bearded [with some exceptions (Burns, 1981b)], ringed, and ribbon seals, haul out only on ice. They may follow and use retreating ice over long distances. Others reside in areas where the ice disappears seasonally. This independence of ice is pronounced in the ribbon seal (Burns, 1981a). The walrus and the spotted seal, although dependent on ice for reproduction, may haul out on land during the ice-free season.

Habitat Features

Physical characteristics of a habitat could influence the way in which a pinniped contacts oil. Habitat features will be examined in the context of seven major categories: sandy shores, rocky shores, fast ice, pack ice, shore leads, polynyas, and oceanic fronts.

Oil on a *sandy shore* is readily rubbed into the pelage of a pinniped. Balancing that, I have observed on Sable Island, Nova Scotia, that clean sand can cleanse oil from pelage by adsorbance and abrasion. On open coasts, sand can also "process" beached oil rather rapidly by adsorption and dispersion, and often by burying it. Temporary burial may extend the period of potential exposure of pinnipeds, but presumably lessens its intensity.

Completely smooth *rocky shores* might receive only a thin coating of oil during a major spill, though depressions and tide pools could accumulate near-smothering amounts. However, crevices and finer scale relief on such shores may sequester oil away from contact by pinnipeds. In less dynamic environments, oil on rocky shores may persist to form a hard, tarry coating.

Fast ice rarely extends beyond headlands; its distribution in Arctic waters in North America has been mapped by McLaren (1958c), Smith and Rigby (1981), Finley *et al.* (1983), and Frost *et al.* (1985). The way in which oil is incorporated into fast ice and subsequently appears in meltpools on the surface has been outlined by Neff (this volume, Chapter 1). Some pinnipeds, notably the ringed seal, depend on fast ice at the mouths of bays and inlets (Kingsley *et al.,* 1985) for a breeding habitat; other northern species such as the bearded seal and walrus only occasionally maintain breathing holes in fast ice (Stirling *et al.,* 1981). There has been some concern that oil may concentrate in these breathing holes (Johnson, 1983; Engelhardt, 1985), yet unless the oil were released directly under the ice or swept under by strong currents, it would not likely spread beyond the barrier presented by the edge of the ice. Instead, oil would accumulate in leads and tide cracks that penetrate into frozen bays and inlets.

Pack ice forms annually to fill virtually all seasonally open water from the Bering Sea, across the Canadian Arctic, to the coasts of Atlantic

Table 3-4
Social Organization of North American Pinnipeds[a]

Species	Breeding structures	Nonbreeding haulouts	Behavior at sea
Walrus	Extensive mobile aquatic leks of displaying adult males attracting estrous females from pack ice.	On land, ice. Segregation of females, adult and subadult males.	Gregarious, groupings as in nonbreeding haulouts.
Steller's sea lion	Stable male territories on land before peak of arrival of gregarious females within them.	Often daily on land, gregarious, sexes and ages mixed.	Often gregarious, possible cooperative foraging.
California sea lion	Labile or stable male territories on land after peak arrival of gregarious females.	Often daily on land, gregarious, sexes and ages mixed.	Often gregarious, possible cooperative feeding.
Northern fur seal	Stable male territories on land before arrival of gregarious females. Some restraint by males.	Rare (pathological?)	Nongregarious, casual associations.
Guadalupe fur seal	Females in male territories on land.	Gregarious on land. Daily (?)	(?)
Bearded seal	Solitary females and pups on ice. Females later attracted by "singing" males in under-ice territories (?) or leks (?).	Nongregarious and opportunistic on pack ice.	Nongregarious, casual associations.

Hooded seal	Scattered in extensive "patches" on pack ice.	Large "molting patches" of females guarded by sequentially polygynous male.	Nongregarious, casual associations?
Ringed seal	Scattered females with pupping lairs in fast (sometimes pack) ice. Mating within under-ice male territories covering one or more lairs (?).	Nongregarious during molt and casually later, on ice.	Nongregarious, casual associations.
Spotted seal	Scattered females pup on pack ice. Female guarded by male on mating in water. Sequentially polygynous (?).	Nongregarious (?) molt on ice? Casual groups on land.	Nongregarious, casual associations.
Harbor seal	Scattered females pup on land, mating in underwater male territories (?) off shorelines.	Solitary or "vigilance groups" on rocks, etc.	Nongregarious, casual associations.
Harp seal	Females with pups, scattered or in large "patches" on pack ice, mating with males in under-ice territories (?).	Large molting assemblages near breeding ice. Casual on ice later, usually in groups.	Gregarious, herd behavior in migration.
Ribbon seal	Scattered females with pups on pack ice, mating with males in under-ice territories (?).	Nongregarious, casually in groups on ice during molting, opportunistically later.	Nongregarious, casual associations.

[a] These summaries are categorical, from a number of general sources and do not include many qualifications and exceptions.

Canada. Old or multi-year ice, which is often very thick, predominates in the Arctic Ocean, drifting into the Bering Sea and Canadian Arctic Archipelago. The southern limits of pack ice in the Bering Sea are sketched on Figs. 3–9, 3–11 and 3–12. In eastern Canada, heavy drift ice of the Labrador current is augmented further south by the outpouring of the Gulf of St. Lawrence, from which extensive ice sometimes (as in spring 1987) reaches Sable Island and Halifax, Nova Scotia. Ice from the Bay of Fundy region rarely intrudes significantly on pinniped habitat.

More than half the pinniped species in North America reproduce on pack ice and use it in other ways (see Tables 3–2 and 3–4). Pack ice concentrates oil between the floes (Ayers *et al.,* 1974) where it could reach smothering levels. This might be especially so in thick, old pack ice. Wind, currents, and proximity to land will also influence the distribution and concentration of oil within the pack, and at the floe edge, where pinnipeds often reside (McLaren, 1958a). In the Bering Sea, spotted and ribbon seals are concentrated along the southern ice front during the winter (Burns, 1978). During migration and summer, young harp seals may actively seek pack ice (Sergeant, 1976), and thus they frequently occur along ice edges, perhaps to exploit prey (Wells and Percy, 1985).

Two other ice configurations influence the distribution and activities of pinnipeds: *shore leads* and *polynyas*. Common to both are enhanced productivity because of ice-edge effects (Dunbar, 1981) and the propensity to accumulate or confine oil. Also, both settings are likely to be used as lanes for shipping oil.

Depending on tides, winds, and season, *shore leads* may open to varying extents at the boundary between fast ice and pack ice. Some are sufficiently recurrent and persistent in winter to qualify as polynyas. Many shore leads in the Canadian Arctic are described and discussed by Smith and Rigby (1981) and Stirling *et al.* (1981). The great spring shore lead off northwest Alaska is depicted by Davis and Thomson (1984). The distribution of polynyas in the Canadian Arctic is mapped and analyzed by Smith and Rigby (1981) and Stirling (1981), and their pattern along the Alaskan coast is depicted in Shapiro and Burns (1975).

Whenever and wherever there is any loosening of the pack ice, pinnipeds literally take advantage of the breathing space. Polynyas have important wintering concentrations of pinnipeds, especially bearded seals, immature ringed seals, walruses (Stirling *et al.,* 1981) and, in the eastern Canadian Arctic, harbor seals (Mansfield, 1967b). The most important shore leads may be recurrent ones that facilitate traditional migration by pinnipeds into summering areas. The open-water lead off northwest Alaska is a major migration corridor for walruses and ringed, spotted, and bearded seals entering the Chukchi sea in spring (Davis andThomson,

1984). Shore leads in Hudson Strait may serve the same function for walruses and harp seals (Stirling *et al.*, 1981).

Oceanic fronts occur at the boundary between stable and vertically mixed water masses (LeFèvre, 1986). It has long been recognized that upwelling water can bring nutrients to phytoplankton in surface waters. It is now understood more generally that when waters are too stable, nutrients become exhausted, whereas with too much vertical mixing, phytoplankton are denied sufficient residence time in sunlit waters. Thus ideal circumstances for high production develop seasonally when and where mixed waters begin to stabilize, and also where there are persistent fronts. The latter may occur at the margins of major oceanic currents, where river plumes contact coastal waters, where currents round headlands, at the edges of underwater banks and most notably along margins of continental shelves. There is some dispute about the extent to which these mixed waters physically concentrate zooplankton as opposed to enhancing its phytoplankton food (LeFèvre 1986). However, there is no doubt that both zooplankton and the pelagic (and larval benthic) fishes that feed on it are abundant in such frontal zones.

Among North American pinnipeds, only the northern fur seal seems closely tied to such large-scale frontal production systems. This species is most common during winter along the edges of the continental shelf from British Columbia to California (Fiscus, 1978) where prevailing westerlies cause upwelling of deep water. The ribbon seal may also take advantage of upwelling along the southern edge of the Bering Sea shelf in summer (Burns, 1981a). It is possible that harp and hooded seals follow the Labrador shelf edge during their seasonal migrations, but the former at least is taken by inshore hunters on the southward migration (Sergeant, 1965). They have also been seen in schools along the edge of ice off Labrador, where they may take advantage of upwelling.

On an ephemeral or local scale, upwelling and downwelling may occur in estuaries with currents or tidal mixing, off points, along small reefs and ridges, or in any area in response to the Langmuir circulation of winds. At sea, areas of upwelling (marked by slicks) and downwelling (marked by flotsam) are often frequented by seabirds and mammals (Buckley *et al.*, 1979). In upwellings, some pinnipeds (e.g., ringed and harp seals) may feed directly on macrozooplankton (euphausiids, hyperiid amphipods) brought to the surface. Others may use the fishes that feed on such macrozooplankton or that "shelter" under seaweed and flotsam which gather in downwelling zones. Local upwelling would act to disperse surface oil, but there may be some concentration in the downwelling zones. Fresh oil and old tar balls are known to accumulate in such flotsam (LeFèvre, 1986), posing risk of ingestion by foraging pinnipeds.

Movements and Migration Patterns

Almost all pinnipeds show regular movements to and from traditional areas or sites (see Table 3–2; Figs. 3–1 to 3–18). In fact, some return repeatedly to precise locations (Pitcher and Calkins, 1983). Even relatively sedentary species that haul out on land make daily forays to feed at sea. Others make coherent, lengthy migrations. All such movements complicate assessment of the probabilities that they may encounter oil.

Harbor seals (Boulva and McLaren, 1979) and females and young of two west coast sea lion species (Fiscus and Baines, 1966) may have regular, once-a-day foraging schedules off their hauling-out sites. Others that undertake more extensive seasonal movements may show such daily forays at certain times of the year. Lactating otariid females make periodic forays to feeding grounds lasting some days (Gentry and Kooyman, 1986).

Some species or populations of pinnipeds (see Table 3–2) show seasonal movements, but not large-scale migrations, from their coastal breeding or haul-out areas to feeding grounds further offshore. A recent study of the California sea lion in the Southern California Bight (Bonnell and Ford, 1987) found about 18% of the population at sea around the breeding islands during the breeding season, compared with about 54% a month later. Most gray seals in eastern Canada are at sea from summer through early winter (Mansfield and Beck, 1977). Adult ringed seals in the eastern Canadian Arctic inhabit fast ice in winter and move offshore in summer. Immatures spend the winter offshore and move into the ice to molt in spring (McLaren, 1958a). Animals offshore are highly dispersed, and large numbers are not likely to contact a spill at sea.

Some populations that disperse offshore seasonally have considerable latitudinal movement that qualifies as migration. Harbor seals seem to move from Maritime Canada to southern New England during winter (Rosenfeld and George, 1985). On the west coast, there are major northward movements of northern elephant seals (Condit and Le Boeuf, 1981) and male California and Steller's sea lions (Mate, 1977) after the breeding seasons. Alaskan populations of Steller's sea lions may move south in winter (Schusterman, 1981).

Some populations of Arctic pinnipeds show seasonal migrations related to ice conditions. During autumn many immature ringed seals appear to vacate the western Canadian Arctic and spend the winter in less icebound parts of the northern Bering Sea (Burns and Eley, 1977; Smith, 1987). Bering Sea stocks of walruses and bearded seals move to the Chukchi Sea in summer (Figs. 3–8 and 3–11). While migrating, these populations traverse a variety of environments where oil might be encountered. Except for the large herds of walrus that may move synchronously through areas subject to oil exploration (Fig. 3–8), these migrations do not appear to concentrate large fractions of the total population of any particular species in localized areas. In general, such populations appear to be less at risk than are sedentary ones, given equal probabilities of an oil spill in any given area.

Large-scale, long-distance, synchronous migrations are undertaken by three North American pinnipeds. Northern fur seals move through oil-exploration areas in the southeastern Bering Sea (Fig. 3–13); harp and hooded seals migrate through potential oil-production areas off Labrador (Figs. 3–3 and 3–4). The harp seal might be particularly vulnerable because it migrates in large groups (Sergeant, 1965).

Habitat Aspects of Reproduction

Characteristics of the habitat where pinnipeds mate and produce and nurture their young may lead them to encounter oil. Pinnipeds are born on solid substrates with ready access to the sea; it is precisely at such boundaries that oil is likely to be spilled and to accumulate. Species or populations that reproduce on offshore islands (e.g., northern fur seals) or pack ice (spotted and ribbon seals) might be less subject to casual encounters with oil than those favoring inshore sheltered localities (harbor and ringed seals). Young pinnipeds, such as gray seals, born and nurtured away from the water's edge (Boness and James, 1979) are less likely to encounter oil than are harbor seals produced on small, sometimes tidally covered bars (Lawson and Renouf, 1987). Young harp seals born near the leading margins of the ice pack (Sergeant, 1976) are more likely to encounter oil than are young hooded seals in the thicker, older floes (Bowen *et al.*, 1987b).

Population Size and Habitat Quality

The rate of recovery of a population from a catastrophic event might depend on how close the population was to the environmental carrying capacity at the time of the event. Other than the effects of crowding in colonies of highly polygynous species, the density-dependent mechanisms involved in the regulation of pinniped populations are poorly understood (McLaren and Smith, 1985). There are hints of mechanisms involving resource use in a few cases. Evidence suggests that the stable or declining population of Pacific walruses consumes approximately the net productivity of its preferred benthic prey from the Bering-Chukchi shelf (Fay, 1982). The declines in recent years of the Steller's sea lion in California (Cooper and Stewart, 1982) and Alaska (Merrick *et al.*, 1987) and of the northern fur seal on the Pribilofs (Fowler, 1982) are not understood, but can be taken as *prima facie* evidence for deterioration of their habitats, whether due to natural changes in their food base, interactions with fisheries, or competition from other pinnipeds.

From the apparently higher frequency of pathological conditions among pinnipeds from the Bering Sea compared with those from the Gulf of Alaska, Fay *et al.* (1979) conclude "that life in the Gulf may be less precarious than in the pack ice, or that populations in the Gulf are in better equilibrium with their environment than are those in the Bering Sea." The Beaufort Sea may be an even more marginal environment, judging from large-scale, long-term secular variations in reproductive rate and body condition in ringed seals, the only pinniped that lives there in significant numbers all year. As noted by Geraci and Smith (1976), during "poorer" years "the effects of an environmental disturbance [on ringed seals] would presumably be more widespread, affecting entire year classes and weakened segments within the population."

Social Organization and Behavior

Among pinnipeds, social organization is not so elaborate as it is in cetaceans, and pinniped groups are often site-dependent and ephemeral. Nevertheless, bonds among individual pinnipeds may lead to multiple exposures to oil.

Although pinnipeds do exhibit elements of courtship display [even group solicitations by females (Heath, 1985)] and pair bonding, precopulation gatherings of land-breeding polygynous species seem to develop more through site fidelity and cohesiveness among females [e.g., California sea lion (Odell, 1981); northern elephant seal (Schusterman, 1981)] than through sexual attachment to individual males. Females would therefore be unlikely to remain in a hazardous situation because of pair-bonding.

Bonding between mothers and their pups is vital when females must discriminate among masses of young (Bartholomew, 1959) or when following behavior is important (Lawson and Renouf, 1987). Even the mother ringed seal, which might only need to recognize a fixed birth site, presumably has the usual mammalian bonding with its dependent pup. The bond is broken during weaning by departure of the female from the breeding site (most species) or by increasing indifference to the pup (Lawson and Renouf, 1987). It is possible that an oiled pup would be unrecognizable to its mother and be prematurely abandoned. A more serious problem might arise if females, as the demographically most important segment of the population, were loathe to abandon pups when threatened by major oil pollution.

Sea lions sometimes nurse older individuals along with the current year's pup (Odell, 1981; Schusterman, 1981). Francis and Heath (1985) determined that most of these older sucklings are female, which might compound the seriousness of an oil spill at a breeding colony.

Nonreproductive Groups

Many pinniped species are gregarious at times other than during breeding. Time spent on ice or ashore saves energy and permits restorative sleep. Even species that form unstructured groups may gain some advantage in surveillance (Krieber and Barrette, 1984). Other species have socially structured groups that benefit individuals. For example, Harestad and Fisher (1975) found that, in a nonpupping colony, adult male Steller's sea lions maintained tranquil areas within which females were free from harassment by subadult males. Disruption of such arrangements could increase energy loss and stress individuals.

Normally "solitary" species such as bearded, ringed, and harbor seals form noninteractive groups at sea, usually in response to localized resources. Without cohesive ties, they might disperse from a spill. Less predictable would be the reaction of harp seals (Sergeant, 1971) and sea lions (Fiscus and Baines, 1966) which dive and swim synchronously, in a display of group cohesion.

Walruses are among the most gregarious of mammals. They occur as small groups at sea, and haul out in groups of up to several thousand. After suckling for two years, weaned calves form unisexual herds, particularly in the Pacific walrus (Fay, 1982). Such segregation allows for selective exposure of females, which determine the reproductive potential of the herd.

Reproductive Behavior

In some species, breeding males posture and fight tenaciously to maintain status or territory. The behavior persists over long periods of time at the expense of feeding. The minimum cost of such reproductive effort is debilitation. Moreover, fighting among highly polygynous species that breed on land (Le Boeuf, 1974) and in the water (Boulva and McLaren, 1979) can result in severe wounding. These elements combine to heighten the level of stress, and thereby affect an animal's reaction to an environmental disturbance.

The period devoted to nurturing young varies greatly among species (Fig. 3–19), from as little as 4 days in the hooded seal (Bowen *et al.*, 1985) to a year or more in the walrus and sea lion. Females with shorter nurturing periods are less restrained to the breeding habitats, and because of efficient transfer of nutrients are more robust at the time of weaning (Bowen *et al.*, 1987a). Such females should be more resilient to oil pollution.

Attendance pattern of nursing females will determine the frequency of exposure to oil on the water or at the ice edge. Some female phocids, for example, the northern elephant seal (Le Boeuf, 1982) and some land-breeding gray seals (Boness and James, 1979), attend their young throughout lactation, thus lessening the risk of repeated exposure. Otariid females, in contrast, make extensive foraging trips to sea during lactation (Gentry and Kooyman, 1986). Harbor seals and most ice-breeding species may return to the water several times a day between suckling bouts (Oftedal *et al.*, 1987). The opportunity for repeated contact with oil would be greater in these animals.

Circumstances surrounding weaning affect survivorship of pups. Sea lions (Odell, 1981; Pitcher and Calkins, 1981) and walruses (Fay, 1982) continue to suckle their young after they have learned to eat other foods. Phocids and the northern fur seal are left to develop their own skills in foraging and consequently have higher mortality rates during this critical period. Oil may compound the difficulty in establishing foraging patterns if it impairs mobility or sensory capability of the pups, redistributes suitable prey, or results in exclusion from favorable areas.

Grooming

Pinnipeds scratch themselves vigorously with their flippers, but do not seem to mouth or lick themselves. Although mammalian mothers routinely lick and mouth-groom their young, this seems to be almost unrecorded among pinnipeds [one exception is the California sea lion in the Galapagos (Eibl-Eibesfeldt, 1956)]. Steller's sea lions may lift and carry their pups by mouth (Schusterman, 1981). Play among young pinnipeds may involve the mouthing of beach debris (Schusterman, 1981). None of these behaviors is likely to contribute to ingestion of significant quantities of oil.

Commensalism with Humans

Like other animals, pinnipeds can become tame when unmolested. Thus harbor seals are reappearing in numbers in harbors in eastern Canada and the west coast. Shaughnessy and Chapman (1984) document the dependence of Cape fur seals on food around Cape Town docks and their increased vulnerability to harbor pollutants.

Demography

A significant effect of oil pollution should be measurable as a population change. Therefore it is important to consider the demography of pinnipeds as a guide to understanding or even predicting such changes.

Life History Parameters

Pinnipeds have many attributes of K-selected species (Eberhardt, 1977). That is, they are late-maturing, slow-reproducing, long-lived animals that can potentially exist in numbers close to the limits set by resources. Females do not mature until at least 3 years old, produce at most 1 young per year (twins rarely), and live well over 20 years (Table 3–5). There are exceptions. In some species, population limits may be set by mortality of young in dense breeding aggregations as a byproduct of sexual selection. This in turn can select for accelerated maturation rate and reduced length of adult life (McLaren, 1967; Reiter, 1984).

There are important differences among species in age of maturity, age-specific fertility rate, and lifespan. However, there are no simple allometric or biogeographic rules relating these parameters to body size (Tables 3–3 and 3–5). Thus, at one extreme females of the northern elephant seal, the largest and one of the more polygynous species, may produce first offspring when less than 4 years old, and annually thereafter for a rather limited lifespan. Northern fur seals, equally polygynous, live longer and do not mature as quickly, perhaps because of the more elaborate requirements of mothering. Sea lions, also highly polygynous but with lengthier periods of parental investment by females (Pitcher and Calkins, 1981), mature later still and live even longer. McLaren (1967) thought that the late maturation and long lives of such "solitary" species as ringed and harbor seals were related to their need for considerable learning about specific environments and circumstances to ensure successful breeding. Mass ice breeders such as harp and ribbon seals, with less need for such knowledge, mature earlier and have shorter lives. The walrus, and to a lesser extent the bearded seal, are interestingly anomalous: females are very late maturing, yet not very long lived (see Table 3–5). There are records of much older male walruses (34 to 38 years in Krylov, 1970; 30+ years in Mansfield, 1958). Is it possible that exploitation of benthic resources or life on the arctic pack ice are particularly stressful, especially to females? In this context, it is noteworthy that the walrus has a unique biennial cycle of reproduction, driven by exceptionally long devotion to the young (Fay, 1982). This, along with late maturity, would make the walrus demographically the most vulnerable of all pinnipeds to population catastrophes.

Table 3-5
Reproductive Parameters of Female North American Pinnipeds

Species	Geographical region	Mean age first young[a]	Adult pregnancy rate[b]	Maximum age[c]	Sources
Walrus	Bering–Chukchi seas	7.2–7.8	0.44	28	Fay (1982); Krylov (1966, 1970)
	E. Canadian Arctic	8.5	0.34	20+	Mansfield (1958)
Steller's sea lion	Gulf of Alaska	5.9	0.87	30	Calkins and Pitcher (1983)
California sea lion	California, Mexico	ca. 8	ca. 1.0	31	Lluch (1969); Maser *et al.* (1981)
Northern fur seal	E. North Pacific	4.8	0.85	26	Lander (1981); York (1983)
Northern elephant seal	California	3.7	0.97	17+	Le Boeuf and Reiter (1988)
Bearded seal	Bering–Chukchi seas	6.4	0.88	23	Burns and Frost (1983)
Hooded seal	Newfoundland	3.8	0.95	30+	Oritsland (1975); Oristland and Benjaminsen (1975); Kapel (1981)
Ringed seal	E. Canadian Arctic	7.9–8.1	0.85–0.93	43	McLaren (1958a); Smith (1973)
	Beaufort Sea	7.7	0.88	36+	Smith (1987)
	Bering–Chukchi seas	7.9–8.3	0.93	29	Fedoseev (1965); Burns and Eley (1977)
Spotted seal	Bering–Okhotsk seas	5.1+	ca. 1.0	35	Tikhomirov (1966, 1968)
Harbor seal	Atlantic Canada	5.2	0.95	32	Boulva and McLaren (1979)
	Alaska	6.2	ca. 1.0	36	Pitcher and Calkins (1983)
Harp seal	Atlantic Canada	4.4	0.94	30+	Bowen *et al.* (1981); Nazarenko and Timoshenko (1974)
Ribbon seal	Bering Sea	3.5–4.0	0.95	26	Burns (1971); Shustov (1965); Tikhomirov (1966)
Gray seal	Atlantic Canada	5.0	0.86	44	Mansfield (1978), Mansfield and Beck (1977)

[a] Where two values are given, they come from the two sources; a + sign indicates that the estimate is based on ovulations rather than pregnancies.
[b] Circa 1.0 implies that most females give birth annually, but that success rates have not been established.
[c] + Indicates that age shown is a minimum estimate.

The Possibility of Local Extinction

The large populations (Table 3–1) and wide geographic distributions (Figs. 3–1 to 3–18) of most pinnipeds make the risk of extinction by oil pollution negligible. However, local populations may be less secure. Ultimately, the life history parameters of a species will determine its fate.

In general, adults of such long-lived animals can withstand short-term environmental disturbances. Furthermore, the loss of a year's offspring may be of little consequence in a population with many overlapping generations. However, *sustained* decreases in survival or fertility rates can lead to inexorable population declines. In late-maturing, low-fertility species, decreases in adult survival rate can be more dangerous than a comparable decrease in fertility rate or equivalent decrease in survival rate of young. This is exemplified using the gray seal as a model species (Fig. 3–20). In a population well below equilibrium, a doubling of mortality will thwart population increase. The same result can only be achieved by reducing fertility rates by about 25% of normal. However, suspected

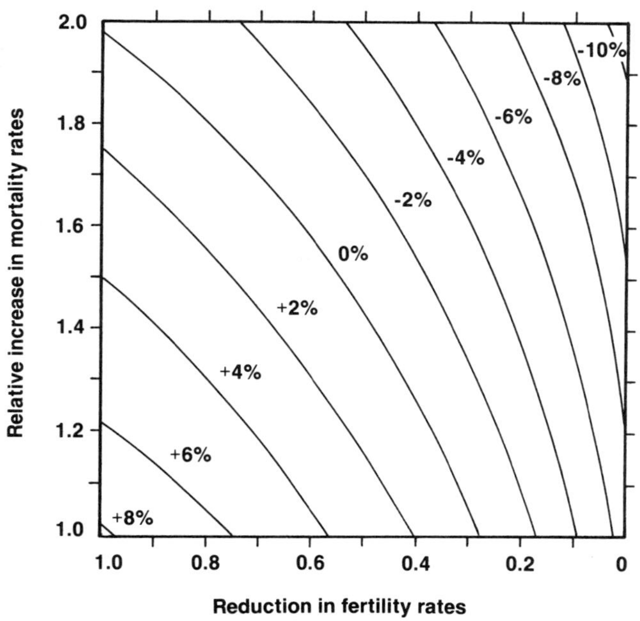

Figure 3-20
Theoretical population response to changes in mortality and fertility rates. The gray seal has been used as the model species, a population well below equilibrium increasing at a rate of 8% per year. If mortality rate doubles or fertility rate decreases by about 25% of maximum, there will be no net increase in population size.

pollution-related population declines of harbor seals (Reijnders, 1986), and ringed and gray seals (Bergman and Olsson, 1986) in Europe, have implicated impaired reproduction rather than increased mortality.

A different kind of threat comes from the remote chance of catastrophic accident with direct, acute mortalities. This could be particularly serious in small, localized populations, already close to the level where "chance" might settle their fate. The role of chance in extinction has long attracted theoretical enquiry. Earlier models assumed that population birth and death rates were on average equal, but with variation among individuals. Under these conditions, extinction is inevitable, although large populations with reasonably small variances may persist for a long time. Recently it has become clear that it is not the "built-in" variance among individuals but variance in environmental conditions that leads rapidly to dangerously low populations.

Though pinnipeds do show attributes of K-selected species, McLaren and Smith (1985) argued that pinnipeds of extreme environments in particular may be considerably influenced by density-independent environmental factors. Thus the ringed seal in the Canadian Beaufort Sea showed a substantial decline in numbers, body condition, and reproductive success during the years 1972–1976 (Smith, 1987). Coincident long-term variations in ringed seal densities have been noted along the Alaskan coast (Frost *et al.,* 1985). Though not fully understood, such population changes may be caused by variation in the marine "climate" and attendant changes in ice cover and food supply.

Clearly, pollution can contribute to the variance of environments in which pinnipeds live. According to Johnson (1983), the accumulation of small environmental perturbations in the Arctic is capable of generating "noise" in energy flow paths, increasing variability, and possibly eliminating important stocks, among which he singles out the walrus as particularly vulnerable. His rather abstract view resists quantitative treatment at present. However, there have been some recent advances relating environmental variance to the probability of local extinctions. Strebel (1985) demonstrated formally that there is a greater probability of extinction when there is resonance between species generation length and the average interval between environmental (or resource) fluctuations. In other words, there is great resilience of long-lived species like pinnipeds to normal seasonal or other short-term fluctuations. However, long-term, lower-amplitude fluctuations, whether natural or anthropogenic, might be more "attuned" to the life cycle characteristics of pinnipeds. Goodman (1987a) has shown that, with purely individual variation, expected persistence time of a population increases as the power of the assumed population "ceiling," whereas with purely environmental variation the time increases somewhat less than linearly with the ceiling. This is a formal analysis of the common sense view that, if some environmental event is bad for the whole population, large numbers will not be a protection.

Another determinant of persistence of a population is the extent of immigration, which obviously can counteract local tendencies toward extinction. Goodman (1987b) demonstrates theoretically that, as might be expected, scattered subpopulations, each experiencing its own environmental variation, with sufficiently high interchanges will persist longer than a single, isolated population with the same overall ceiling. Species of pinnipeds with well-developed homing or patrophilic tendencies would thus be more prone to local extinction than would be those that wander unpredictably.

Another recurrent concern about rare and diminishing populations is the reduction of genetic variation that occurs from inbreeding, with its supposed consequences for individual fitness. Earlier work seemed to indicate that the northern elephant seal had indeed already experienced such genetic depauperation as a result of earlier overexploitation (Bonnell and Selander, 1974). It was speculated that this would make them more vulnerable to such influences as pollution. However, it is now clear that pinnipeds are naturally homozygous at most enzyme loci (Lidicker *et al.*, 1981), so that reduced genetic variance is probably not a potentially dangerous consequence of local population reductions. This conclusion may be revised with current technology for studying nuclear and mitochondrial DNA.

An excellent consideration of the above theoretical possibilities for pinnipeds is found in the study of the isolated ringed seals in Lake Saimaa, Finland (Jarvinen and Varvio, 1986). The population is clearly endangered, with only some 130–150 individuals remaining. Thus, it is tempting to apply models of stochastic extinction. However, as Jarvinen and Varvio (1986) argue, "when one traces the history of endangered or extinct species, it is not stochastic extinction that one typically finds, but rather a tragedy of persecution and habitat deterioration," and "the absolute rule of chance in a small population is only the final stage of a long process leading to extinction." They conclude that the reduction of the seal stock is not a result of stochastic fluctuations, even amplified, in environmental conditions, but rather stems from *persistent* pressures. Previously high hunting mortality in adults has been succeeded by heightened mortality of pups due to artificial water level changes, entanglement in fishing gear, and possibly to pollutants.

The theoretical literature on the demography of extinction thus offers few insights into processes that could lead to local extinction of populations of seals because of impacts from oil. It does seem, however,

Table 3-6
Direct Estimates of Rates of Increase of Unexploited or Lightly Exploited Pinniped Populations Thought To Be Recovering at High Rates Following Earlier Overexploitation, and Believed To Be Uncomplicated by Immigration

Species	Locality	Years	Increase[a]	Sources
California sea lion	S. California	1927–1946	1.091	Chapman (1981)
Northern fur seal	Pribilof Islands, Alaska	1912–1924	1.085	Chapman (1981)
Antarctic fur seal	South Georgia	1958–1972	1.168	Payne (1977)
Subantarctic fur seal	Gough Island	1955–1977	1.149	Bester (1980)
Northern elephant seal	California, various islands	1964–1981	1.146–1.793	Cooper and Stewart (1983)
Gray seal	Farnes and Outer Hebrides, Britain	1950–1976	1.067–1.073	Summers (1978)

[a] Rates of increase are multiples per year.

that catastrophes are likely to be less important than small, but sustained, reductions in survival and fertility. Chronic oil pollution could contribute to such effects.

Population Recovery After Catastrophe

If we take as a premise that an oil pollution catastrophe has produced a large kill of pinnipeds, it is important to consider the rate at which the population is restored to its original level. Indeed, rate of return to "normality" is often an explicit component of environmental impact statements. The rate of recovery depends on the species and circumstances; the estimates available are for populations of pinnipeds under conditions that probably produce close to the maximum values (Table 3–6). Clearly pinniped populations could recover at rates of 7–18% per year, provided they were well below equilibrium levels before the catastrophe. Those closer to equilibrium should show a less enhanced rate of increase following substantial reduction of numbers (Table 3–7). A population already limited by the carrying capacity of the environment may not be capable of producing an immediate population response. Some species, such as the northern fur seal and Steller's sea lion (see Table 3–1), are indeed declining locally, presumably because the carrying capacity of their environments is in some way deteriorating. For these, a catastrophic kill might not be followed by population recovery.

The rate of recovery of a seal population will also depend on the segments of the population that are killed. Table 3–7 illustrates, using the British gray seal as a demographic model, that loss of a year's offspring is

Table 3-7
Recovery Times of Hypothetical Seal Populations after Various Catastrophic Reductions, Assumed to Occur Immediately After the Young Are Born

Scenario	Population removed (%)	Time to 95% recovery (yr)
Population initially at equilibrium size		
All newborns killed	23	2
Equal proportions all age classes killed	23	37
Equal proportions mature age classes killed	23	43
Population initially at one-half equilibrium size		
All newborns killed	20	1
Equal proportions of all age classes killed	23	6
Equal proportions of mature age classes killed	23	7

[a] Population is based on the same model gray seal population used for Fig. 3-20.

more quickly compensated than is an equivalent loss of all age groups from the population, or particularly of adults. Yet even the long recovery time for populations at equilibrium before a catastrophe (Table 3–7) might be optimistic if individuals in crowded environments fail to respond to population reduction.

References

Ayers, R. C., Jr., Johns, H. O., and Glaseser, J. L. (1974). Oil spills in the Arctic Ocean: Extent of spreading and possibility of large-scale thermal effects. *Science* **186,** 842–845.

Bartholomew, G. A. (1959). Mother-young relations and the maturation of pup behaviour in the Alaska fur seal. *Anim. Behav.* **7,** 163–171.

Bayer, R. D. (1981). California sea lion in the Yaquina River Estuary, Oregon. *Murrelet* **62,** 56–59.

Beloborodov, A. G., and Potelev, V. A. (1966). First period of life of the hooded seal in the Greenland Sea. *In* "Proceedings of the Third All-Union Conference Marine Mammals," pp. 7–8. Izd. "Nauka," Vladivostok (in Russian).

Bergman, A., and Olsson, M. (1986). Pathology of Baltic gray seal and ringed seal females with special reference to adrenocortical hyperplasia: Is environmental pollution the cause of a widely distributed disease syndrome? *Finn. Game Res.* **44,** 47–62.

Bester, M. N. (1980). Population increase in the Amsterdam Island fur seal *Arctocephalus tropicalis* at Gough Island. *S. Afr. J. Zool.* **15,** 229–234.

Bigg, M. A. (1969a). The harbour seal in British Columbia. *Bull. Fish. Res. Board Can.* **172,** 1–33.

Bigg, M. A. (1969b). Clines in the pupping season of the harbour seal, *Phoca vitulina*. *J. Fish. Res. Board Can.* **26,** 449–455.

Bigg, M. A. (1981). Harbour seal *Phoca vitulina* Linnaeus, 1758 and *Phoca largha* Pallas, 1811. *In* "Handbook of Marine Mammals" (S. H. Ridgway and R. J. Harrison, eds.), Vol. 2, pp. 1–27. Academic Press, London.

Boness, D. J., and James, H. (1979). Reproductive behaviour of the gray seal (*Halichoerus grypus*) on Sable Island, Nova Scotia. *J. Zool.* **188,** 477–500.

Boness, D. J., Bowen, W. D., and Oftedal, O.T. (1988). Evidence of polygyny from spatial patterns of hooded seals (*Cystophora cristata*). *Can. J. Zool.* **66,** 703–706.

Bonnell, M. L., and Ford, R. G. (1987). California sea lion distributions: A statistical analysis of aerial transect data. *J. Wildl. Manage.* **51,** 13–20.

Bonnell, M. L., and Selander, R. K. (1974). Elephant seals: Genetic variation and near extinction. *Science* **184,** 908–909.

Boulva, J., and McLaren, I. A. (1979). Biology of the harbor seal, *Phoca vitulina*, in eastern Canada. *Bull. Fish. Res. Board Can.* **200,** 1–24.

Bowen, W. D., Capstick, C. K., and Sergeant, D. E. (1981). Temporal changes in the reproductive potential of female harp seals (*Pagophilus groenlandicus*). *Can. J. Fish. Aquat. Sci.* **38,** 495–503.

Bowen, W. D., Boness, D. J., and Oftedal, O. T. (1987a). Mass transfer from mother to pup and subsequent mass loss by the weaned pup in the hooded seal, *Cystophora cristata*. *Can. J. Zool.* **65,** 1–8.

Bowen, W. D., Myers, R. A., and Hay, K. (1987b). Abundance estimation of a dispersed, dynamic population: Hooded seals (*Cystophora cristata*) in the Northwest Atlantic. *Can. J. Fish. Aquat. Sci.* **44,** 282–295.

Bowen, W. D., Oftedal, O. T., and Boness, D. J. (1985). Birth to weaning in 4 days: Remarkable growth in the hooded seal. *Can. J. Zool.* **63,** 2481–2486.

Braham, H. W., Oliver, G. W., Fowler, C., Frost, K., Fay, F., Cowles, C., Costa, D., Schneider, K., and Calkins, D. (1982). Marine mammals. *In* "The St. George Basin Environment and Possible Consequences of Offshore Oil and Gas Development" (M. J. Hameedi, ed.), Proc. Synth. Meet, 1981, pp. 55–81. U.S. Govt. Printing Office, Washington, D.C.

Buckley, J. R., Gammelsrod, T., Johannessen, J. A., Johannessen, O. M., and Roed, L. P. (1979). Upwelling: Oceanic structure at the edge of the arctic pack in winter. *Science* **203,** 165–167.

Burns, J. J. (1971). Biology of the ribbon seal, *Histriophoca fasciata*, in the Bering Sea. *Sci. Alaska* **22,** 135.

Burns, J. J. (1978). Ice seals. *In* "Marine Mammals of Eastern North Pacific and Arctic Waters"(D. Haley, ed.), pp. 193–205. Pacific Search Press, Seattle, Washington.

Burns, J. J. (1981a). Ribbon seal *Phoca fasciata* Zimmermann, 1783. *In* "Handbook of Marine Mammals" (S. H. Ridgway and R. J. Harrison, eds.), Vol. 2, pp. 89–109. Academic Press, New York.

Burns, J. J. (1981b). Bearded seal, *Erignathus barbatus*, Erxleben, 1777. *In* "Handbook of Marine Mammals" (S. H. Ridgway and R. J. Harrison, eds.), Vol. 2, pp. 145–170. Academic Press, New York.

Burns, J. J., and Eley, T. J. (1977). The natural history and ecology of the bearded seal (*Erignathus barbatus*) and the ringed seal (*Phoca hispida*). *In* "Environmental Assessment of the Alaskan Continental Shelf," Vol. 1, pp. 226–302. U.S. Dept. of Commerce, Natl. Oceanic Atmos. Admin., Washington, D. C.

Burns, J. J., and Frost, K. J. (1983). The natural history and ecology of the bearded seal, *Erignathus barbatus*. *In* "Environmental Assessment of the Alaskan Continental Shelf," Final Rep., Outer Continental Shelf Environ. Assess. Program, Vol. 19, pp. 311–392. U.S. Dept. of Commerce, Natl. Oceanic Atmos. Admin., Washington, D.C.

Burns, J. J., and Gol'tsev, V. N. (1984). Comparative biology of harbor seals, *Phoca vitulina* Linnaeus, 1758, of the Commander, Aleutian, and Pribilof Islands. *NOAA Tech. Rep., NMFS* **12,** 17–24.

Burns, J. J., Ray, G. C., Fay, F. H., and Shauhgnessy, P. D. (1972). Adoption of a strange pup by the ice-inhabiting harbor seal, *Phoca vitulina largha*. *J. Mammal.* **53,** 594–598.

Calambokidis, J., Everitt, R. D., Cubbage, J., and Carter, S. D. (1979). Harbor seal census for the inland waters of Washington, 1977-78. *Murrelet* **60,** 110–111.

Calkins, D. G., and Pitcher, K. W. (1983). Population assessment, ecology and trophic relationships of Steller sea lions in the Gulf of Alaska. *In* "Environmental Assessment of the Alaskan Continental Shelf," Final Rep., Outer Continental Shelf Environ. Assess. Program, Vol. 19, pp. 445–546. U.S. Dept. of Commerce, Natl. Oceanic Atmos. Admin., Washington, D.C.

Chapman, D. G. (1981). Evaluation of marine mammal population models. *In* "Dynamics of Large Mammal Populations" (C. W. Fowler and T. D. Smith, eds.), pp. 279–296. Wiley, New York.

Condit, R., and Le Boeuf, B. J. (1981). Pelagic distribution of the northern elephant seal, *Mirounga angustirostris*. *Abstr. 4th Bienn. Conf. Biol. Mar. Mammals*, San Francisco.

Cooper, C. F., and Stewart, B. S. (1982). "The Perils of Success: Implications of Increasing Marine Mammal Populations in the Southern California Bight," Ocean Stud. Symp., Calif. Coastal Comm., Monterey.

Cooper, C. F., and Stewart, B. S. (1983). Demography of northern elephant seals, 1911-1982. *Science* **219,** 969-971.

Davis, R. A., and Thomson, D. H. (1984). Marine mammals. *In* "The Barrow Arch Environment and Possible Consequences of Planned Offshore. Oil and Gas Development" (J.C. Truett, ed.), Outer Continental Shelf Environ. Assess. Program, pp. 47–79. Natl. Oceanic Atmos. Admin., Ocean Assess. Div., Anchorage, Alaska.

Davis, R. A., Finley, K. W., and Richardson, W. J. (1980). "The Present Status and Future Management of Arctic Marine Mammals in Canada," Rep. Sci. Advisory Board N.W.T., Yellowknife, N.W.T.

DeLong, R. L. (1978). Northern elephant seal. *In* "Marine Mammals of Eastern North Pacific and Arctic Waters" (D. Haley, ed.), pp.207–211. Pacific Search Press, Seattle, Washington.

DeMaster, D. P., Miller, D. J., Goodmand, D., DeLong, R. L., and Stewart, B. S. (1982). Assessment of California Sea Lion Fishery Interactions. *Trans. North Am. Wildl. Natl. Resour. Conf.* **47,** 253–264.

Dunbar, M. J. (1981). Physical causes and biological significance of polynyas and other open water in sea ice. *Can. Wildl. Serv., Occas. Pap.* **45,** 29–43.

Eberhardt, L. L. (1977). Optimal policies for conservation of large mammals with special reference to marine ecosystems. *Biol. Conserv.* **4,** 205–212.

Eibl-Eibesfeldt, I. (1956). The Otariidae of the Galapagos Islands. *J. Mammal.* **37,** 459.

Eisenberg, J. F., and Kleiman, D. G., eds. (1983). "Advances in the Study of Mammalian Behavior," Spec. Publ. Am. Soc. Mammal.

Engelhardt, F. R. (1985). Effects of petroleum on marine mammals. *In* "Petroleum Effects in the Arctic Environment" (F. R. Engelhardt, ed.), pp. 217–243. Elsevier Appl. Sci., London.

Estes, J. A., and Gol'tsev, V. N. (1984). Abundance and distribution of the Pacific walrus, *Odobenus rosmarus divergens*: Results of the first Soviet-American joint aerial survey, autumn 1975. *NOAA Tech. Rep., NMFS* **12,** 67–76.

Everitt, R. D., and Beach, R. J. (1982). Marine mammal-fisheries interactions in Oregon and Washington: An overview. *Trans. North Am. Wildl. Nat. Resour. Conf.* **47,** 265–277.

Everitt, R. D., and Braham, H. W. (1980). Aerial survey of Pacific harbor seals in the southwestern Bering Sea. *Northwest Sci.* **54,** 281–288.

Fay, F. H. (1982). "Ecology and Biology of the Pacific Walrus, *Odobenus rosmarus divergens* Illiger," North Am. Fauna No. 74. U.S. Dept. of the Interior, U.S. Fish Wildl. Serv., Washington, D.C.

Fay, F. H., Dieterich, R. A., Schults, L. M., Murray, N. K., Hoover, A., and Kelly, B. P. (1979). "Morbidity and Mortality of Marine Mammals," Annu. Rep., Contract No. 03-5-022-56. U.S. Dept. of Commerce, Natl. Oceanic Atmos. Admin., Environmental Assessment of the Alaskan Outer Continental Shelf, Washington, D.C.

Fedoseev, G. A. (1965). Comparative characteristics of the ringed seal populations in the coastal waters of the Chukostk Peninsula. Izv. *TINRO* **59,** 194–212 (in Russian).

Fedoseev, G. A. (1973). Morpho-ecological characteristics of ribbon seal populations and the basis for protection of its stocks. *Tr. TINRO* **86,** 158–177 (Fish. Res. Board Can. Transl. Ser. 3365).

Finley, K. J., Miller, G. W., Davis, R.A., and Koski, W. R. (1983). A distinctive large breeding population of ringed seals (*Phoca hispida*) inhabiting the Baffin Bay pack ice. *Arctic* **36,** 162–173.

Fiscus, C. H. (1961). Growth in the Steller's sea lion. *J. Mammal.* **42,** 218–223.

Fiscus, C. H. (1978). Northern fur seal. *In* "Marine Mammals of Eastern North Pacific and Arctic Waters" (D. Haley, ed.), pp. 153–159. Pacific Search Press, Seattle, Washington.

Fiscus, C. H., and Baines, G. A. (1966). Food and feeding behavior of Steller and California sea lions. *J. Mammal.* **47,** 195–200.

Fleischer, L. A. (1978). Guadalupe fur seal. *In* "Marine Mammals of Eastern North Pacific and Arctic Waters" (D. Haley, ed), pp. 161–165. Pacific Search Press, Seattle, Washington.

Fowler, C. W. (1982). Interactions of northern fur seals and commercial fisheries. *Trans. North Am. Wildl. Nat. Resour. Conf.* **47,** 278–292.

Fowler, C. W., and Smith, T. D., eds. (1981). "Dynamics of Large Mammal Populations." Wiley, New York.

Francis, J. M., and Heath, C. B. (1985). Duration of maternal care in the California sea lion—bias by sex. *Abstr., 6th Bienn. Conf. Biol. Mar. Mammals,* Vancouver, B. C.

Frost, K. J., and Lowry, L. F. (1984). Trophic relationships of vertebrate consumers in the Alaskan Beaufort Sea. *In* "The Alaskan Beaufort Sea: Ecosystems and Environments" (P. W. Barnes, D. M. Schell, and E. Remnitz, eds.), pp. 381–401. Academic Press, New York.

Frost, K. J., Lowry, L. F., and Burns, J. J. (1983). Distribution of marine mammals in the coastal zone of the eastern Chukchi Sea during summer and autumn. *In* "Environmental Assessment of the Alaskan Continental Shelf," Final Rep. Outer Continental Shelf Environ. Assess. Program, Vol. 20, pp. 563–650. U.S. Dept. of Commerce, Natl. Oceanic Atmos. Admin., Washington, D.C.

Frost, K. J., Lowry, L. F., and Burns, J. J. (1985). "Ringed Seal Monitoring: Relationships of Distribution, Abundance, and Reproductive Success to Habitat Attributes and Industrial Activities," Interim Rep. Contract No. 84-ABC-00210. Proj. No. RU # 667. Natl. Oceanic Atmos. Admin., Washington, D.C.

Gentry, R. L., and Johnson, J. H. (1981). Predation by sea lions on northern fur seal neonates. *Mammalia* **45,** 423–430.

Gentry, R. L., and Kooyman, G. L., eds. (1986). "Fur Seals. Maternal Strategies on Land and at Sea." Princeton Univ. Press, Princeton, New Jersey.

Geraci, J. R., and Smith, T. G. (1976). Direct and indirect effects of oil on ringed seals (*Phoca hispida*) of the Beaufort Sea. *J. Fish. Res. Board Can.* **33,** 1976–1984.

Gilmartin, W. G., DeLong, R. L., Smith, A. W., Sweeney, J. C., De Lappe, B. W., Riseborough, R. W., Griner, L. A., Dailey, M. D., and Peakall, D. B. (1976). Premature parturition in the California sea lion. *J. Wildl. Dis.* **12,** 104–115.

Goodman, D. (1987a). The demography of chance extinction. *In* "Viable Populations for Conservation (M. E. Soulé, ed.), pp. 11–34. Cambridge Univ. Press, London and New York.

Goodman, D. (1987b). Considerations of stochastic demography in the design and management of biological reserves. *Nat. Resour.* Model **1,** 205–234.

Harestad, A. S., and Fisher, H. D. (1975). Social behavior in a non-pupping colony of Steller sea lions (*Eumetopias jubata*). *Can. J. Zool.* **53,** 1596–1613.

Heath, C.B. (1985). The effects of environment on the breeding system of the California sea lion (*Zalophus californianus*). *Abstr. 6th Bienn. Conf. Biol. Mar. Mammals,* Vancouver, B.C.

Innes, S., Stewart, R. A., and Lavigne, D. M. (1981). Growth in Northwest Atlantic harp seals *Phoca groenlandica*. *J. Zool.* **194,** 11–24.

Jarvinen, J., and Varvio, S.-L. (1986). Proneness to extinction of small populations of seals: Demographic and genetic stochasticity vs. environmental stress. *Finn. Game Res.* **44,** 6–18.

Johnson, L. (1983). Assessment of the effects of oil on arctic marine fish and marine mammals. *Can. Tech. Rep. Fish. Aquat. Sci.* **1200,** 1–15.

Johnson, M.L., Fiscus, C. H., Ostenson, B. T., and Barbour, M. L. (1966). Marine mammals. *In* "Environment of the Cape Thompson Region, Alaska" (N. J. Wilimovsky and J. N. Wolfe, eds.), pp. 877–924. U.S. At. Energy Comm., Oak Ridge, Tennessee.

Kajimura, H. (1984). Opportunistic feeding of the northern fur seal, *Callorhinus ursinus*, in the eastern North Pacific Ocean and eastern Bering Sea. *NOAA, Tech. Rep., NMFS SSRF* **779,** 1–49.

Kapel, F. O. (1981). Studies of hooded seal in Greenland, 1970–1980. Northwest Atlantic Fisheries Organization, Scientific Research Document 81/X1/150, Series N462, 18 pp.

King, J. (1983). "Seals of the World." Cornell Univ. Press, Ithaca, New York.

Kingsley, M. C., Stirling, I., and Calvert, W. (1985). The distribution and abundance of seals in the Canadian High Arctic, 1980-1982. *Can. J. Fish. Aquat. Sci.* **42,** 1189–1210.

Krieber, M., and Barrette, C. (1984). Aggregation behaviour of harbour seals at Florillon National Park, Canada. *J. Anim. Ecol.* **53,** 913–928.

Krylov, V. I. (1966). Sexual maturation in the female Pacific walrus. *Zoologicheskii Zhurnal* **45,** 919–927. [Fisheries Research Board of Canada Translation Series 806].

Krylov, V. I. (1970). Size and weight characteristics as indicators of age and sexual dimorphism in the Pacific walrus. *Mosk. Obshchestr. Isp. Prir. Otd. Biol. Biul.* **5,** 18–24.

Lander, R. H. (1981). A life table and biomass estimate for Alaskan fur seals. *Fish. Res.* **1,** 55–70.

Lavigne, D. M., Innes, S., Worthy, G. A. J., Kovacs, K. M., Schmitz, O. J., and Hickie, J. P. (1986). Metabolic rates of seals and whales. *Can. J. Zool.* **64,** 279–290.

Lawson, J. W., and Renouf, D. (1987). Bonding and weaning in harbor seals. *J. Mammal.* **68,** 445–449.

Le Boeuf, B. J. (1974). Male-male competition and reproductive success in elephant seals. *Am. Zool.* **14,** 163–176.

Le Boeuf, B. J. (1979). Northern elephant seal. *In* "Mammals of the Sea," Vol. 2, Fish. Ser. No. 5, pp. 110–114. Food Agric. Organ. United Nations, Rome.

Le Boeuf, B. J. (1982). The elephant seal. *In* "Problems in Management of Locally Abundant Wild Mammals" (P. A. Jewell, S. Holt, and D. Hart, eds.), pp. 291–301. Academic Press, New York.

Le Boeuf, B. J. and Reiter, J. (1988). Lifetime reproductive success in northern elephant seals. *In* "Reproductive Success: Studies of Individual Variation in Contrasting Breeding Systems" (T. H. Clutton-Brock, eds.), pp. 344–362. Univ. Chicago Press, Chicago.

Le Boeuf, B. J., Aurioles, D., Condit, R., Fox, C., Gisner, R., Romero, R., and Sinsel, F. (1983). Size and distribution of the California sea lion population of Mexico. *Proc. Calif. Acad. Sci.* **43,** 77–85.

Le Boeuf, B. J., Whiting, R. J., and Gantt, R. F. (1972). Perinatal behavior of northern elephant seal females and their young. *Behaviour* **34,** 121–156.

LeFèvre, J. (1986). Aspects of the biology of frontal systems. *Adv. Mar. Biol.* **23,** 163–299.

Lidicker, W. Z., Sage, R. D., and Calkins, D. G. (1981). Biochemical variation in northern sea lions from Alaska. *In* "Mammalian Population Genetics" (M. H. Smith and J. Joule, eds.), pp. 231–241. Univ. of Georgia Press, Athens.

Lluch, B. D. (1969). Crecimento y mortalidad del lobo marino de California *Zalophus californianus*. *An. Esc. Nac. Cienc. Biol.* (*Mexico City*) **18,** 167–189.

Loughlin, T. R., Rugh, D.J., and Fiscus, C. H. (1984). Northern sea lion (*Eumetopias jubatus*) distribution and abundance; 1956-1980. *J. Wildl. Manage.* **48,** 729–740.

Lowry, L. F., and Fay, F. H. (1984). Seal eating by walruses in the Bering and Chukchi Seas. *Polar Biol.* **3,** 11–18.

Lowry, L. F., and Frost, K. J. (1981). Feeding and trophic relationships of phocid seals and walruses in the eastern Bering Sea. *In* "The Eastern Bering Sea Shelf: Oceanography and Resources" (D. W. Wood and J. A. Calder, eds.), Vol. 2, pp. 813–824. Office of Mar. Pollut. Assess. Natl. Oceanic Atmos. Admin., Bur. Land Manage., Washington, D.C.

Lowry, L. F., Frost, K. J., and Burns, J. J. (1980). Variability in the diet of ringed seals, *Phoca hispida*, in Alaska. *Can. J. Fish. Aquat. Sci.* **37,** 2254–2261.

Lucas, Z., and McLaren, I. A. (1988). Apparent predation by gray seals, *Halichoerus grypus*, on seabirds around Sable Island, Nova Scotia. *Can. Field Nat.* **102,** 675–678.

Mansfield, A. W. (1958). The biology of the Atlantic walrus *Odobenus rosmarus rosmarus* (Linnaeus) in the eastern Canadian arctic. *Fish. Res. Board Can. Manuscr. Rep.* **653,** 1–146.

Mansfield, A. W. (1967a). Seals of arctic and eastern Canada. *Bull. Fish. Res. Board Can.* **127,** 1–35.

Mansfield. A. W. (1967b). Distribution of the harbor seal, *Phoca vitulina* Linnaeus, in Canadian Arctic waters. *J. Mammal.* **48,** 249–257.

Mansfield, A. W. (1978). "Reproduction in the Gray Seal *Halichoerus grypus* in Eastern Canada," C.M. 1978/M:13. Int. Counc. Explor. Sea, Mar. Mammal Comm., Charlottenlund, Denmark.

Mansfield, A. W., and Beck, B. (1977). The gray seal in eastern Canada. *Tech. Rep.—Environ. Fish. Mar. Serv. (Can.)* **704,** 1–81.

Mansfield, A. W., and Sergeant, D. E. (1960). Harbour seal population studies. *Fish. Res. Board Can. Arct. Unit Annu. Rep., 1959-1960,* pp. 27–30.

Maser, C., Mate, B. R., Franklin,, J. F., and Dymes, C. T. (1981). Natural history of Oregon coast mammals. U.S. Department of Agriculture, Forestry Service, Technical Report PNW-133.

Mate, B. R. (1977). "Aerial Censusing of Pinnipeds in the Eastern Pacific for Assessment of Population Numbers, Migratory Distributions, Rookery Stability, Breeding Effort, and Recruitment," Final Rep., Contract No. MM5AC001. U.S. Mar. Mammal Comm., Washington, D.C. (U.S. Nat. Tech. Inf. Serv., Springfield, Virginia, No. PN-265 859).

Mate, B. R., and Gentry, R. L. (1979). Northern (Steller) sea lion. *In* "Mammals in the Seas," Vol. 2, Fish. Ser. No. 5. pp. 1–4. Food Agric. Organ. United Nations, Rome.

McGinnis, S. M., and Schusterman, R. J. (1981). Northern elephant seal *Mirounga angustirostris* Gill, 1866. *In* "Handbook of Marine Mammals" (S. H. Ridgway and R. J. Harrison, eds.), Vol. 2, pp. 329–349. Academic Press, New York.

McLaren, I. A. (1958a). The biology of the ringed seal (*Phoca hispida* Schreber) in the eastern Canadian arctic. *Bull. Fish. Res. Board Can.* **118,** 1–97.

McLaren, I. A. (1958b). Some aspects of growth and reproduction of the bearded seal, *Erignathus barbatus* (Erxleben). *J. Fish. Res. Board Can.* **15,** 219–227.

McLaren, I. A. (1958c). The economics of seals in the eastern Canadian arctic. *Fish. Res. Board Can., Arct. Circ.* **1,** 1–94.

McLaren, I. A. (1967). Seals and group selection. *Ecology* **48,** 104–110.

McLaren, I. A., and Smith, T. G. (1985). Population ecology of seals: Retrospective and prospective views. *Mar. Mammal Sci.* **1,** 54–83.

Merrick, R. L., Loughlin, T. R,. and Calkins, D. G. (1987). Decline in abundance of the northern sea lion, *Eumetopias jubatus*, in Alaska, 1955-86. *Fish. Bull.* **85,** 351–365.

Nazarenko, Yu. I., and Timoshenko, Yu. I. (1974). Age structure in the White Sea population of *Pagophoca groenlandica* as an index of efficiency of protective measures. *Zoologicheskii Zhurnal* **53,** 256–262 [in Russian.]

Obee, B. (1984). Steller's sea lion. *Wildl. Rev.* **10**(9), 9–13.

Odell, D. K. (1981). California sea lion *Zalophus californianus* (Lesson, 1828). *In* "Handbook of Marine Mammals" (S. H. Ridgway and R. H. Harrison, eds.), Vol. 1, pp. 67–97. Academic Press, New York, NY.

Oftedal, O. T., Boness, D. J., and Tedman, R. A. 1987. The behavior, physiology, and anatomy of lactation in the pinnipedia. *Curr. Mammal.* **1,** 175–245.

Øritsland, T. (1975). Sexual maturity and reproductive performance of female hooded seals at Newfoundland. *Int. Comm. Northwest Atl. Fish., Res. Bull.* **11,** 37–41.

Øritsland, T., and Benjaminsen, T. (1975). Sex ratio, age composition and mortality of hooded seals at Newfoundland. *Int. Comm. Northwest Atl. Fish., Res. Bull.* **11,** 135–143.

Payne, M. R. (1977). Growth of a fur seal population. *Philos. Trans. R. Soc. London, Ser. B* **279,** 67–79.

Payne, P. M., and Schneider, D. C. (1984). Yearly changes in abundance of harbor seals, *Phoca vitulina*, at a winter haulout site in Massachusetts. *Fish. Bull.* **82,** 440–442.

Pitcher, K. W. (1977). Population productivity and food habits of harbor seals in the Prince William Sound—Copper River Delta area, Alaska. Final Report. U.S. Mar. Mammal Comm. Contract No. NM5AC0011. *U.S. NTIS, PB Rep.* **PB 266935,** 1–36.

Pitcher, K. W., and Calkins, D. G. (1981). Reproductive biology of Steller sea lions in the Gulf of Alaska. *J. Mammal.* **62,** 599–605.

Pitcher, K. W., and Calkins, D. G. (1983). Biology of the harbor seal, *Phoca vitulina richardsi*, in the Gulf of Alaska. *In* "Environmental Assessment of the Alaskan Continental Shelf," Final Rep., Outer Continental Shelf Environ. Assess. Program, Vol. 19, pp. 231–311. U.S. Dept. of Commerce, Natl. Oceanic Atmos. Admin., Washington, D.C.

Pitcher, K. W., and Fay, F. H. (1982). Feeding by Steller sea lions on harbor seals. *Murrelet* **63,** 70–71.

Reeves, R., and Ling, J. K. (1981). Hooded seal *Cystophora cristata* Erxleben, 1777. *In* "Handbook of Marine Mammals" (S. H. Ridgway and R. J. Harrison, eds.), Vol. 2, pp. 171–194. Academic Press, London.

Reijnders, P. J. H. (1986). Reproductive failure in common seals feeding on fish from polluted coastal waters. *Nature (London)* **324,** 456–457.

Reiter, J. (1984). Studies of female competition and reproductive success in the northern elephant seal. Ph.D. Thesis, University of California, Santa Cruz.

Richardson, D. T. (1975). Hooded seal whelps at South Brooksville, Maine. *J. Mammal.* **56,** 698–699.

Roff, D. A., and Bowen, W. D. (1983). Population dynamics of the Northwest Atlantic harp seal (*Phoca groenlandica*). *Can. J. Fish. Aquat. Sci.* **40,** 919–932.

Roff, D. A., and Bowen, W. D. (1986). Further analyses of population trends in the Northwest Atlantic harp seal (*Phoca groenlandica*) from 1967 to 1985. *Can. J. Fish.Aquat. Sci.* **43,** 553–564.

Rosenfeld, M., and George, M. (1985). Migration in harbor seals. *Abst., 6th Bienn. Conf. Biol. Mar. Mammals,* Vancouver B.C.

Schusterman, R. J. (1981). Steller sea lion *Eumetopias jubatus* (Schreber, 1776). *In* "Handbook of Marine Mammals" (S. H. Ridgway and R. J. Harrison, eds.), Vol. 1, pp. 119–141. Academic Press, London.

Sergeant, D. E. (1965). Migrations of harp seals *Pagophilus groenlandicus* (Erxleben) in the Northwest Atlantic. *J. Fish. Res. Board Can.* **22,** 433–464.

Sergeant, D. E. (1971). Migration and orientation in harp seals. *Proc. 7th Ann. Conf. Biol. Sonar Diving Mammals,* Stanford Res. Inst., Menlo Park, California, pp. 123–131.

Sergeant, D. E. (1974). A rediscovered whelping population of hooded seals *Cystophora cristata* Erxleben and its possible relationship to other populations. *Polarforschung* **44,** 1–7.

Sergeant, D. E. (1976). History and present status of populations of harp and hooded seals. *Biol. Conserv.* **10,** 95–118.

Shapiro, L. H,. and Burns, J. J. (1975). "Major Late-Winter Features of Ice in Northern Bering and Chukchi Seas as Determined from Satellite Imagery," Rep. No. UAG R-236 (Sea Grant Rep. No. 75-8). Univ. Alaska Geophys. Inst., Anchorage.

Shaughnessy, P. D., and Chapman, P. (1984). Commensal Cape fur seals in Cape Town docks. *S. Afr. J. Mar. Sci.* **2,** 81–91.

Shustov, A. P. (1965). Some biological features and reproductive rates of the ribbon seal (*Histriophoca fasciata*) in the Bering Sea. *Izv. TINRO* **59,** 183–192 [in Russian].

Shustov, A. P., and Yablokov, A. V. (1967). Comparative morphological characteristics of the harp and ribbon seals. *Tr. PINRO* **21,** 51–59.

Smith, T. G. (1973). Population dynamics of the ringed seal in the eastern Canada Arctic. *Bull. Fish. Res. Board Can.* **181,** 1–55.

Smith, T.G., (1975). Ringed seals in James and Hudson Bay: Population estimate and catch statistics. *Arctic* **28,** 170–182.

Smith, T. G. (1987). The ringed seal, *Phoca hispida*, of the Canadian Western Arctic. *Can. Bull. Fish. Aquat. Sci.* **216,** 1–81.

Smith, M., and Rigby, B. (1981). Distribution of polynyas in the Canadian arctic. *Can. Wildl. Serv., Occas. Pap.* **45,** 7–28.

Spotte, S., and Schneider, J. (1982). Early functional maturity of captive male elephant seals (*Mirounga angustirostris*). *Zoo Biol.* **1,** 355–358.

Stewart, B. S. (1980). Historical and present populations of pinnipeds in the Channel Islands. *In* "Potential Effects of Space Shuttle Sonic Booms on the Biota and Geology of the California Channel Islands: Research Reports" (J. R. Jehl, Jr. and C. F. Cooper, eds.), Tech. Rep. 80-2, pp. 45–98. Center for Marine Studies, San Diego State University, San Diego, California.

Stewart, R. A. E., and Lavigne, D. M. (1980). Neonatal growth of Northwest Atlantic harp seals, *Pagophilus groenlandicus*. *J. Mammal.* **61,** 670–680.

Stirling, I. (1981). Introduction. *Can. Wildl. Serv., Occas. Pap.* **45,** 5–6.

Stirling, I., Archibald, R., and DeMaster, D. (1977). Distribution and abundance of seals in the eastern Beaufort Sea. *J. Fish. Res. Board Can.* **34,** 976–988.

Stirling, I., Cleator, H., and Smith, T. G. (1981). Marine mammals. *Can. Wildl. Serv., Occas. Pap.* **45,** 45–58.

Strebel, D. E. (1985). Environmental fluctuations and extinction—single species. *Theor. Popul. Biol.* **27,** 1–26.

Summers, C. F. (1978). Trends in the size of British gray seal populations. *J. Appl. Ecol.* **15,** 395–400.

Tikhomirov, E. A. (1966). On the reproduction of seals belonging to the family Phocidae in the North Pacific. *Zoologicheskii Zhurnal* **45,** 275–281. [Fisheries Research Board of Canada Translation Series 1889.]

Tikhomirov, E. A. (1968). Body growth and development of reproductive organs of the North Pacific hair seal. *Izv. TINRO* **68,** 216–243 (Israel Program for Sci. Transl., Keter Press, Jerusalem).

Wells, P. G., and Percy, J. A. (1985). Effects of oil on arctic invertebrates. *In* "Petroleum Effects in the Arctic Environment" (F. R. Engelhardt, ed.), pp. 101–156. Elsevier Appl. Sci., London.

Wiig, O. (1985). Morphometric variation in the hooded seal (*Cystophora cristata*). *J. Zool. Ser. A* **206,** 497–508.

York, A. (1983). Average age at first reproduction of the northern fur seal (*Callorhinus ursinus*). *Can. J. Fish. Aquat. Sci.* **40,** 121–127.

Zwanenburg, K., Bowen, W. D., and Sergeant, D. E. (1985). "Assessment of Northwest Atlantic Gray Seal (*Halichoerus grypus*) Pup Production for 1977 to 1984," Res. Doc. 85/67. Can. Dept. Fish. Oceans Sci. Advis. Comm. Atl. Fish. 16 pp.

4

Physiologic and Toxic Effects on Pinnipeds

David J. St. Aubin
Department of Pathology
Ontario Veterinary College
University of Guelph
Guelph, Ontario
N1G 2W1, Canada

Historical Notes

The record of encounters between oil and seals, sea lions, and their relatives spans four decades and comprises at least 29 events, occurring on the coasts of Europe and Antarctica, and on the Pacific, Atlantic, and Arctic coasts of North America. Armed with information from such diverse habitats, we should have little difficulty in predicting the consequences of oil exposure for most pinniped species. Yet the quality of the reports is variable. Early accounts are often vague or brief, and some of the more recent ones, prepared in an emotionally charged atmosphere, are less than objective. Despite these shortcomings, the record clearly indicates that pinnipeds are vulnerable to and may be harmed by oiling.

Oil well blowouts and tanker accidents attract considerable attention, and some of these have claimed a number of pinnipeds. However, more insidious sources such as vessels that intentionally discharge fuel can in fact be as damaging. During the late 1940s, two instances were recorded in which seals were affected by petroleum presumed to have been deliberately released from ships. In the Antarctic, Lillie (1954) observed "half-grown seals covered in a sticky, tarry mess, their eyes bloodshot with irritation." Davies (1949) reported that two seal pups encased in oil drowned when washed out to sea near Ramsay Island off the coast of Wales; other oiled pups were apparently unaffected. Waste oil dumped

Sea Mammals and Oil: Confronting the Risks
Copyright © 1990 by Academic Press, Inc.
All rights of reproduction in any form reserved.

from vessels may account for other incidents of oil fouling (Table 4–1). Generally, reports of these events are based on the recovery of two or three oil-stained animals, with little evidence to show whether the seals were fouled before or after death.

A systematic study was performed following an oil spill along the coast of Wales in 1974 (Davis and Anderson, 1976). Oil from an unknown source fouled gray seal rookeries at the onset of the pupping season, a very sensitive period in their life history. On Skomer Island, 25 pups and 23 adults were wholly or partly oiled. Though pup mortality there was higher than at other less affected sites, there was no significant difference in mortality rate between oiled and clean pups. Necropsy examinations did not reveal gross evidence of ingested petroleum. The only deaths directly attributed to fouling were those of two pups so encased with oil that they drowned when washed off the beach.

Tanker groundings have resulted in major oil spills that have affected pinnipeds. The first such event was the sinking of the *Torrey Canyon* off the coast of England in 1967. Two or three gray seals were observed surfacing in the oil slick, and three oiled animals were recovered dead or dying (Gill *et al.*, 1967); Spooner (1967) reported that as many as 12 seals (species not indicated) had died. In view of the magnitude of the spill and the extent of clean-up activities which involved the use of over 2.5 million gal of relatively toxic dispersants, the immediate impact of this event on pinnipeds was minor.

Subsequently, major spills from the *Arrow*, *Amoco Cadiz*, *Christos Bitas*, *Kurdistan*, *Bahia Paraiso* and *Exxon Valdez* had similar consequences. Harbor and gray seals were fouled in Chedabucto Bay, Nova Scotia, where the tanker *Arrow* released 4 million gal of Bunker C fuel oil in February, 1970 (Fig. 4–1). On Sable Island, 200 km to the south, most of the 50–60 harbor and 100 gray seals occupying the beaches were fouled to some degree. A few animals were thought to have suffocated because vital orifices had been plugged (Anonymous, 1970a). Nine years later, the tanker *Kurdistan* sank in Cabot Strait, 200 km northeast of the wreck of the *Arrow*, and oiled gray and harbor seals were again observed on Sable Island (Parsons *et al.*, 1980) (Fig. 4–1). Investigators found no evidence of mortality or physical impairment even though some seals were heavily oiled. Along the nearby coast of Nova Scotia, however, four dead gray seals and six dead harbor seals were found coated with oil.

Following the sinking of the *Amoco Cadiz* in March 1978, two of four dead gray seals recovered were fouled with oil; autopsies were not performed to confirm the cause of death (Prieur and Hussenot, 1978). In October, 1978, the grounded *Christos Bitas* discharged over 840,000 gal of Iranian crude oil along the coast of Wales in the same area where Davis and

Table 4-1
Reports of Pinnipeds Associated with Oil

Date	Location and source	Oil type and quantity[a]	Species	Impact	Reference
late 1940s	Antarctic; ship discharge	Fuel oil, Quan. ?	Unspecified seals	Bloodshot eyes; surface fouling with tarry oil	Lillie (1954)
1949	Ramsay Island, Wales; source unknown	Fuel oil, Quan. ?	Gray seal	Pups largely unaffected by thick coating of oil. Two fouled pups drowned.	Davies (1949)
Mar. 1967	English Channel; *Torrey Canyon*	Crude oil, 30×10^6 gal	Gray seal	Seals observed surfacing through slick. 3 oiled seals found dead or dying. Up to 12 confirmed deaths.	Gill *et al.* (1967); Spooner (1967)
Jan. 1969	Gulf of St. Lawrence; storage tank	Bunker C, 4000 gal	Harp seal	10–15,000 seals coated. Unspecified number of dead seals recovered.	Warner (1969); Sergeant (1987)
Feb. 1969	Santa Barbara, Calif.; Union Oil well	Crude oil, $> 30 \times 10^6$ gal	Harbor seal, elephant seal, Calif. sea lion	Oiled seals observed on Channel Islands and along mainland coast. Mortalities not conclusively linked to oil.	Le Boeuf (1971); Brownell and Le Boeuf (1971); Simpson (1970)
Nov. 1969	N. Dyfed, Wales; source unknown	Type ?, Quan. ?	Gray seal	14 oiled, dead pups found. No causal relationship.	Anonymous (1970b)

(*Table continues*)

Table 4-1 (*Continued*)

Date	Location and source	Oil type and quantity[a]	Species	Impact	Reference
Feb. 1970	Chedabucto Bay, Sable Island, N.S.; *Arrow*	Bunker C, 4×10^6 gal	Gray seal, harbor seal	50–60 harbor seals and 100 gray seals oiled on Sable Island. 500 oiled seals in Chedabucto Bay. 24 found dead, some with oil in mouth or stomach.	Anonymous (1970a, 1971a)
Feb.–Mar. 1970	Kodiak Island, AK.; ship discharge	Slop oil or oily ballast	Hair seals, sea lions	Est. 500 mammals contacted. No mortalities.	Hess and Trobaugh (1971)
Apr. 1970	Alaska Peninsula; source unknown	Diesel fuel, Quan. ?	Hair seals	400 seals exhibited unusual behavior. No mortalities.	Anonymous (1971b)
Nov. 1970	Farne Islands; source unknown	Type ?, Quan. ?	Gray seal	Yearling seal found with oil-stained pelt and crusting around mouth. Otherwise healthy.	Bonner and Hickling (1971)
Mar. 1972	British Columbia; *Vanlene*	Bunker B, 100,000 gal	Seals	Seal herds in area unaffected.	Environmental Sciences Ltd. (ESL) (1981)
Sept. 1973	Repulse Bay, NWT; ship discharge	Refuse oil, Quan. ?	Ringed seal	Hunters killed 5 oil-covered seals.	Muller-Willie (1974)
1973	Dutch coast; source unknown	Type ?, Quan. ?	Harbor seal	Patches of oil inconclusively associated with skin lesions.	van Haaften (1973)

1974–1979	Cape Town, S.A.; ships and industry	Chronic discharge	Cape fur seal	Fur seals lingering in polluted harbor without obvious effect.	Shaughnessy and Chapman (1984)
Aug. 1974	Straits of Magellan; *Metula*	Crude oil, 14×10^6 gal Bunker C, 1×10^6 gal	S. sea lion, S. Am. fur seal	Sea lions and fur seals in the area apparently unaffected.	Baker *et al.* (1976)
Aug. 1974	Coast of France; source unknown	Fuel oil, Quan. ?	Harbor seal, gray seal	Oil in intestine of 1 harbor seal. 3 oiled gray seals, 1 ingested oil.	Duguy and Babin (1975)
Sept. 1974	Pembrokeshire, Wales; source unknown	Type ?, Quan. ?	Gray seal	2 heavily oiled pups drowned when washed off beach. 25 pups and 23 adults fouled.	Davis and Anderson (1976)
Jan. 1975	Ireland; *African Zodiac*	Bunker C, 1.1×10^6 gal	Seals	Seals in the area were apparently unaffected.	ESL (1981)
Aug. 1977	Greenland; *USNS Potomac*	Bunker C, 1×10^6 gal	Ringed seal, other seals	16 oiled seals observed 1 month after spill.	Grose *et al.* (1979)
Mar. 1978	France; *Amoco Cadiz*	Crude oil, 60×10^6 gal	Gray seal	2 of 4 dead seals coated with oil. No causal relationship.	Prieur and Hussenot (1978)
May 1978	Great Yarmouth; *UK Eleni V*	Heavy fuel, 1×10^6 gal	Seals	20 oiled seals observed.	ESL (1981)
Oct. 1978	South Wales; *Christos Bitas*	Crude oil, 840,000 gal	Seals	Mortality of 16 of 23 oiled.	Bourne (1979)
Dec. 1978	Shetland Is.; Scotland, *Esso Bernicia*	Bunker C, 370,000 gal	Seals	Some seals oiled. No deaths reported.	Anderson (1981)
Feb. 1979	Latvia; *Antonio Gramsci*	Crude oil, 36,500 gal	Seal	One seal killed by oil.	ESL (1981)

(*Table continues*)

Table 4-1 (*Continued*)

Date	Location and source	Oil type and quantity[a]	Species	Impact	Reference
Mar. 1979	Cabot Str., N.S.; *Kurdistan*	Bunker C, 2.1×10^6 gal	Gray seal, harbor seal	At least 4 gray and 6 harbor seals found dead coated with oil. No causal relationship. Oiled seals on Sable Is.	D. F. Parsons *et al.* (1980); D. F. Marston (personal communication)
Nov. 1979	Pribilof Is., AK.; *F/V Ryuyo Maru*	Fuel oil, 290,000 gal	Northern fur seal	Some oiled dead pups found. Causal relationship not demonstrated.	Reiter (1981)
Feb. 1984	Sable Is., N.S.; Well blow out	Gas condensate	Gray seal	4 oiled seals observed on Sable Is. No mortalities.	Anonymous (1984)
Jan. 1989	Anvers Is., Antarctica; *Bahia Paraiso*	Diesel fuel, 233,000 gal	Crabeater and elephant seals, southern fur seals	2 crabeater seals affected. Elephant seals and fur seals oiled but unharmed.	Anonymous (1989); T. De Laca (personal communication)
Mar. 1989	Prince William Sound, Alaska; *Exxon Valdez*	Crude oil, 11×10^6 gal	Harbor seal, fur seals, and Steller's sea lion	Seals observed swimming in oil. 31 harbor seal, 2 fur seal, and 14 sea lion carcasses with some oil.	D. G. Calkins (personal communication)

[a] All volumes converted to U.S. gallons and rounded. 1 barrel = 159 L = 42 gal. 1 ton = 278 gal of bunker oil, 300 gal of crude oil, or 332 gal of diesel fuel.

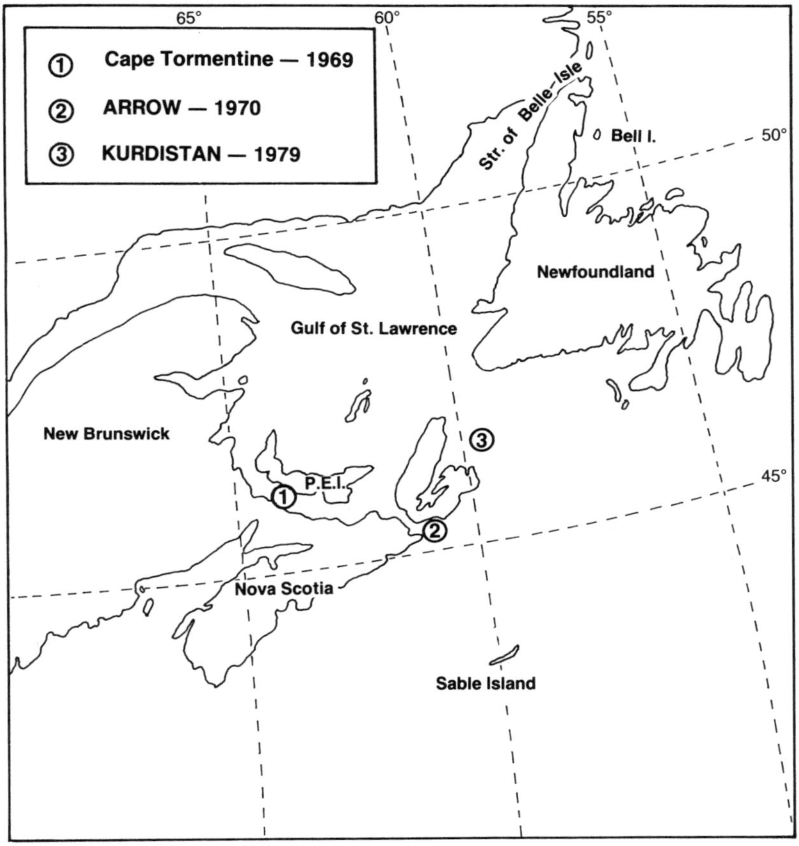

Figure 4-1
Gray, harbor, and harp seals were oiled after three spills along the Atlantic coast of Canada.

Anderson had investigated a spill 5 years earlier. Reports that 16 gray seal pups died were not confirmed, and subsequent necropsy examinations of a limited number of specimens were inconclusive (Bourne, 1979).

The supply ship *Bahia Paraiso* released most of its 233,000 gal cargo of diesel fuel when it ran aground near the tip of the Antarctic peninsula (Anonymous, 1989). Apparently disoriented crabeater seals were observed surfacing in the slick, and a few oiled animals hauled out nearby. Two subadults were particularly listless and had blood-tinged nasal discharge and feces (T. De Lacà, personal communication). The seals were so debilitated that they were attacked by scavenging seabirds.

Their fate was not determined. Oil-fouled elephant seals and southern fur seals were apparently unaffected.

The March 1989 grounding of the *Exxon Valdez* released 11 million gal of crude oil into the habitat of harbor seals, northern fur seals, and Steller's sea lions in Prince William Sound, Alaska. Large numbers of animals were observed swimming in the slick (D. G. Calkins, personal communication). The spill occurred 1 month before the onset of pupping season for harbor seals and fouled rookeries where pups subsequently became contaminated; most of the 31 oiled harbor seals recovered were pups. The oiled carcasses of 2 fur seals and 14 sea lions were also collected for examination, but the results of this investigation have not been released at this time.

Oiled seals have been observed following the breakup of at least seven other vessels, with mortalities reported in two of the events. An unspecified number of oil-fouled dead fur seal pups were recovered in the Pribilof Islands after the grounding of the fishing vessel *F/V Ryuyo Maru* (Reiter, 1981), and a single seal (species not identified) died presumably because of contacting crude oil spilled from the *Antonio Gramsci* in the Baltic Sea (Environmental Sciences Ltd., 1981). In neither instance was a detailed examination performed on an animal.

No incident involving marine mammals and spilled oil has sparked as much controversy as the blowout of Union Oil's A-21 well in the Santa Barbara Channel off the coast of California(Fig. 4–2) on January 28, 1969 (Easton, 1972). The magnitude and duration of the spill, and its occurrence near a densely populated coast, drew national attention for several months. Residents who perceived the spill as an irreversible blow to the economic and esthetic value of the coastal environment confronted industry and federal representatives as conflicting reports of the extent of the spill and its impact on marine life were presented and challenged. Studies that found minimal effects were dismissed by the public as inadequate, whereas media reports were often overstated and sensational and found little favor with the scientific community. From this large body of diverse and often conflicting documentation, some attempt has been made to evaluate the impact on pinnipeds.

Along the mainland coast, there was little evidence of harm. Ten days after the blowout, the Santa Barbara *News-Press* published a photograph of an oiled harbor seal that had been "rescued" and presented for cleaning. By February 21, *Time* magazine (Anonymous, 1969a) reported that six dead seals had washed up on California beaches. Seven more were recovered during the following month.

Most of the attention and controversy centered around rookeries on the Channel Islands—San Miguel, San Nicholas, Santa Cruz, and

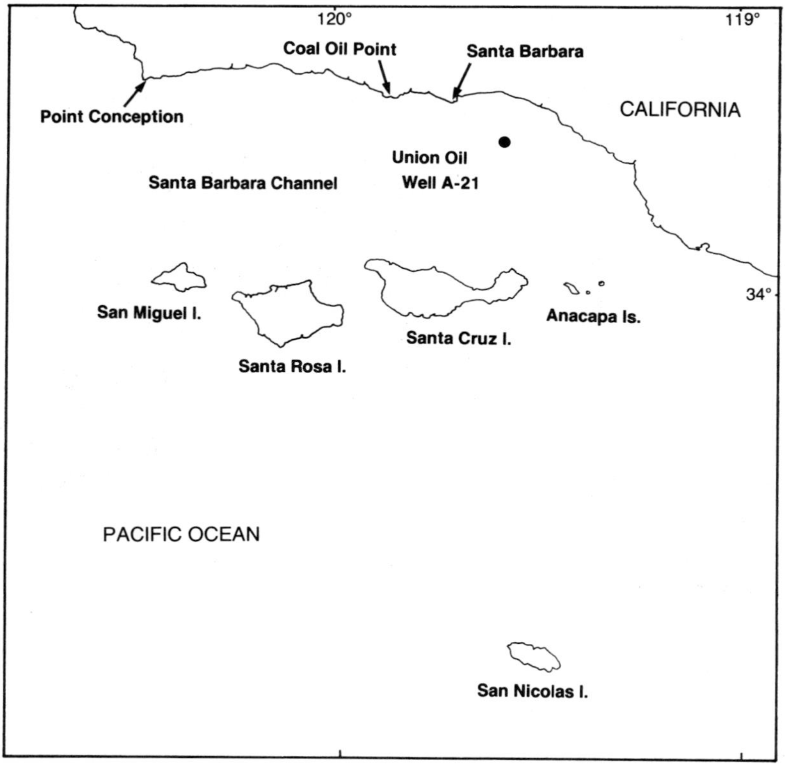

Figure 4-2
In 1969 oil from Union Oil's well A-21 reached islands occupied by California sea lions and elephant seals in the Santa Barbara Channel.

Anacapa (Fig. 4–2)—which are occupied throughout the year by several species of pinnipeds. From early February until the end of June, the islands were surveyed on at least 15 occasions by groups representing the California Department of Fish and Game, the University of California (Santa Barbara and Santa Cruz), the Defenders of Wildlife, the Santa Barbara Museum of Natural History, and various press agencies, including *Time* magazine. Their observations and interpretations of the events form the basis of the controversy.

The principal species of pinnipeds on the islands are California sea lions and elephant seals, both of which use the islands for breeding. Elephant seal pups are born in early January, and most were weaned when oil reached the islands one week after the blowout. On March 17, Le Boeuf (1971) observed over 100 elephant seal pups coated with oil, sand, and

A heavy oiled northern elephant seal on San Miguel Island following the January 1969 blowout of a Union Oil platform in the Santa Barbara Channel. Photo by B. J. Le Boeuf.

detritus. Fifty-eight of these were tagged, along with an equal number of unoiled animals. During the next 15 months, 40% of the oiled group and 25% of the controls were resighted. Conclusions about survivorship were tenuous, since there was no assurance that seals that were clean when tagged did not subsequently encounter oil. Later, Le Boeuf and Peterson reported that they had also observed four dead elephant seals lying in pools of oil (Anonymous, 1969b). During the first two weeks in April, Simpson and Gilmartin (1970) surveyed the rookery on four occasions and found three dead seals, with no evidence of oil contamination. No hydrocarbon residues were detected in tissues collected from two of these, nor in blood samples from two live seals.

The effect of oil on California sea lions was the subject of particular concern and disagreement. Though peak pupping season for the sea lions does not occur until late May through June (McLaren, this volume, Chapter 3), observers were confounded by an apparently large number of premature births, beginning in February. Counts ranged from 25 in late March (Brownell and Le Boeuf, 1971) to 200 in mid-May (Snell, 1969), just before the onset of the normal parturition season. Arguments arose over whether the incidence of abortions was representative of normal mortality.

Sand and detritus readily adhered to the oil-fouled surfaces of a northern elephant seal pup. Photo by B. J. Le Boeuf.

No hydrocarbon residues were detected in tissue samples from aborted fetuses (Simpson and Gilmartin, 1970; Simpson, 1970), and the association between pup mortality and the oil spill remained circumstantial. Subsequent investigations of premature births among sea lions showed that the incidence observed in 1969 was not unusual. DeLong *et al.* (1973) counted 242 dead pups on April 25, 1970, and 348 on May 18, 1971, on San Miguel Island; Odell (1970) reported 442 on San Nicholas Island between January 17 and May 3, 1970. More recent studies have shown possible associations of premature births with infectious organisms and organochlorine residues (Gilmartin *et al.*, 1976); perhaps the stress of oil exposure had an additive effect. To summarize, there was no clear evidence directly implicating oil as the principal cause of mortality among sea lion pups.

Concurrent with the Santa Barbara spill, a much less publicized incident in Canadian waters had a far greater impact on pinnipeds. A ruptured storage tank on Cape Tormentine, New Brunswick, discharged

4000 gal of Bunker C oil into the Gulf of St. Lawrence (see Fig. 4–2), fouling the sea ice where harp seals were about to pup. During March and early April, observers reported 10,000–15,000 oil-fouled adults and pups, all so heavily contaminated that they were almost unrecognizable (Sergeant, 1987). Hundreds of pups were tagged as part of an annual population study by Canadian Fisheries biologists, and the relatively high tag return during this season suggested an increase in mortality rate. An unspecified number of dead seals were found on the ice floes, in the water, and along the beaches of Bell Island (Warner, 1969). Oil-fouled pups migrated normally towards the Strait of Belle Isle, despite the heavy coating which observers felt would impede swimming (Sergeant, 1987). A telling clue to their condition, however, was that the pups were easily taken weeks later by seal hunters who reported that the animals uncharacteristically would not leave the ice floes (Sergeant, 1987). This incident combined the worst possible conditions of viscous residual oil in an ice-infested, cold-water environment at a particularly vulnerable period in the life of a pinniped.

Detection and Avoidance of Oil

Pinnipeds are physiologically and anatomically well-equipped to detect oil, though no study has addressed this issue experimentally. They have reasonably acute vision (Nachtigall, 1986), particularly underwater over a wide range of light intensities. A large pupil and extensive summation of rod-type receptors facilitate discrimination under such low light conditions (Jamieson and Fisher, 1972), and a well-developed tapetum further enhances this ability. Pinnipeds can feed at night (Renouf *et al.*, 1980) or at great depths. In ice-covered seas, where ambient light is further reduced, Arctic seals readily travel between breathing holes, demonstrating an ability to locate discontinuities at the surface.

Pinnipeds appear to have a good sense of smell. They use olfaction to identify their young (Sandegren, 1970; Fogden, 1971; Renouf *et al.*, 1983), and seals basking on ice floes test the air while maintaining their vigil against predators. One would expect that their olfactory sense is keen enough to detect hydrocarbon vapors.

It is unlikely that any pinniped has an acoustic sense as sophisticated as that found in some odontocetes. Studies on the ability of seals to echolocate (Renouf *et al.*, 1980; Renouf and Davis, 1982) have been challenged (Wartzok *et al.*, 1984). In the absence of other cues, they may take maximum advantage of their hearing; apparently blind pinnipeds can survive for some time in the wild (King, 1983). However this provides no insight as to whether a seal can detect oil beyond its ability to see or smell it.

Pinnipeds would seem to be able to detect oil because some are able to avoid it. Here again, the evidence is not conclusive. Most of the several thousand resident gray and harbor seals apparently left Chedabucto Bay, Nova Scotia, after the grounding of the *Arrow* (Mansfield, 1970), though their response may have been as much to the marked increase in human activity as to the oil itself. Harbor seals temporarily abandoned Yell Sound in the Shetland Islands, Scotland, when Bunker C oil from the *Esso Bernicia* escaped from containment booms (Anderson, 1981). The seals returned several weeks later, after most of the oil had beached. Native hunters from Wainwright, Alaska, reported that an ongoing fuel spill which resulted in a 5-km-long slick was responsible for low numbers of seals in the area (Cowles *et al.*, 1981), though their absence may have related more to changes in food distribution and abundance. Relatively few seals were observed in the vicinity of oil spilled from the *USNS Potomac*, due more to the lack of sea ice than to the presence of oil (Grose *et al.*, 1979).

The evidence for oil avoidance behavior in pinnipeds is contradicted by observations of seals, sea lions, and fur seals swimming in the midst of oil slicks. After the *Torrey Canyon* spill, two gray seals were seen "deliberately" diving and surfacing in a patch of oil (Spooner, 1967). Seals and sea lions did not abandon rookeries fouled after the oil well blowout in the Santa Barbara Channel, and showed no reluctance to enter oiled waters surrounding Anacapa and Santa Cruz islands when disturbed by personnel from the California Department of Fish and Game (Battelle Memorial Institute, 1969). In Alaskan waters, fur seals were apparently indifferent to small slicks of oil and mousse released from the *F/V Ryuyo Maru* (Reiter, 1981), and harbor seals and Steller's sea lions surfaced repeatedly in the spill from the *Exxon Valdez* (D. G. Calkins, personal communication). Under confined laboratory conditions, three ringed seals placed in a pen containing seawater with a 1-cm-thick slick made no attempt to use a haul-out platform. However, the platform was not provided until the animals had been in the pen for 20 minutes, and by then the seals may have been too excited or disoriented to notice its presence (Geraci and Smith, 1976a).

An intriguing account of the response of pinnipeds to oil comes from observations made over a 5-year period in Table Bay Harbour, Cape Town, South Africa (Shaughnessy and Chapman, 1984). There, in inner reaches of the harbor where petroleum residues accumulate at the surface and in the water column, Cape fur seals regularly come to feed on fish inadvertently discharged from trawlers. The authors could not determine how long individuals remained within the harbor, or whether more seals would have been present if pollution levels were lower. Nevertheless, their observations demonstrate that a pinniped intent on a ready meal is not discouraged by the mere presence of oil.

Behavioral Effects

Oil spills could have a disruptive effect on individuals or populations by interfering with normal behavior patterns. Of particular concern is the effect on maternal behavior. Pinnipeds appear to rely on scent to establish a mother–pup bond (Sandegren, 1970; Fogden, 1971). Oil-coated pups may not be recognizable to their mothers as suggested by the study on San Miguel Island, where media representatives reported that female sea lions were biting and tossing pups that might have been their own (McMillan, 1969). J. Bennett of the University of California, Santa Barbara, Museum of Zoology, observed in a limited survey that females appeared to ignore oiled pups that attempted to suckle (Santa Barbara *News-Press,* June 29, 1969). In a systematic study of nursing behavior in oiled gray seals, Davis and Anderson (1976) concluded that mother–pup interactions were normal, though oiled pups had lower peak weights at weaning. Frequent attempts by the researchers to clean the pups likely disturbed nursing behavior more than did the oil itself. Earlier, Davies (1949) had noted that gray seals continued to nurse pups that were heavily contaminated with fuel oil.

Cape fur seals feeding within the heavily polluted Table Bay Harbour were observed to behave normally (Shaughnessy and Chapman, 1984). An unusual behavior, noted on four occasions during the 5-year observation period, was an uncharacteristic swimming posture in which a seal carried its head, neck and trunk above water for extended periods of time. No conclusion could be made as to whether these animals were responding to irritating substances in the water.

Fouled seals may be reluctant to enter the water, according to observers in the Gulf of St. Lawrence (Sergeant, 1987) and Alaska (Anonymous, 1971b). The seals in Alaskan waters had "a glazed look in their eyes," and were possibly disoriented after contacting a spill of light diesel fuel. The young harp seals in the Gulf, though weaned and mature enough to swim, were impeded by a heavy coating of Bunker C fuel oil which stuck their flippers to their sides.

Surface Contact

Eyes and Mucous Membranes

The most sensitive tissues exposed to the environment are the mucous membranes that surround the eyes and line the oral cavity, respiratory surfaces, and anal and urogenital orifices. Petroleum hydrocarbons, particularly volatile aromatics and short-chain fractions, are irritating to these and other delicate tissues such as the cornea. People exposed to hydrocarbon vapors report irritation of eyes and respiratory epithelium at relatively low concentrations (Davis *et al.*, 1960).

Ringed seals experimentally placed in crude oil-covered water showed a similar reaction (Smith and Geraci, 1975). Within minutes after exposure to oil, the seals began to lacrimate profusely and eventually had difficulty keeping their eyes open. By 24 hours, they developed severe conjunctivitis, swollen nictitating membranes, and corneal abrasions and ulcers. The inflammation subsided soon after the seals were placed in clean water. It is reasonable to assume that continued exposure could have resulted in permanent damage.

Similar effects have been observed in the natural setting (Lillie, 1954). In addition, pinnipeds contacting highly weathered petroleum face persistent contamination of periocular tissues because of the tenacity of these viscous substances. Further compounding the problem is the tendency for tar to entrap debris, such as sand and sticks, to the point where some animals may have difficulty opening their eyes (Anonymous, 1969a). We can surmise that pinnipeds, with their relatively large, often protruding eyes would be particularly vulnerable to such effects.

Integument

Though not as sensitive as mucous membranes, epidermis can also be damaged by petroleum, particularly the low molecular weight fractions (Walsh *et al.*, 1974; Hansbrough *et al.*, 1985). These components remove protective lipids from the skin surface, penetrate between epidermal cells, disrupt cellular membranes, and elicit an inflammatory response in the dermis (Lupulescu *et al.*, 1973). Necrotic epidermis is generally sloughed, leaving ulcers. In humans, this can occur after contact for less than an hour (Klauder and Brille, 1947; Tagami and Ogino, 1973).

Despite the potential for cutaneous damage, such lesions have rarely been noted on oil-fouled seals. Van Haaften (1973) supposed that skin lesions on harbor seals recovered in Dutch waters resulted from contact with oil, but acknowledged that similar lesions occur without evidence of oil contamination. In a controlled experiment, Geraci and Smith (1976b) found no indication of skin damage in ringed seals immersed in oil-covered water for 24 hours. Perhaps their dense wettable underfur prevented much of the oil from contacting the epidermis. Without evidence, we can only speculate that there would be a greater opportunity for contact and perhaps greater risk in species with relatively sparse pelage, such as the California sea lion and the walrus.

Of greater concern is the potential effect of surface fouling on thermoregulation. The marine environment is particularly demanding energetically, due to low temperatures and high specific heat of the medium. Pinnipeds are well adapted to withstand immersion in near-freezing seas, using anatomic features and physiological mechanisms to maintain core body temperature. For insulation, they rely on a thick layer of subcutaneous fat (blubber), dense fur, or both. In an *in vitro* experiment, Kooyman *et al.* (1976, 1977) compared thermal conductance through pelts from northern fur seals, bearded and Weddell seals and California sea lions. The dense pelage of the northern fur seal provided the greatest resistance to heat transfer, whereas specimens from bearded seals and the sea lion had much higher conductance values.

Fur is an effective thermal barrier because it traps air and repels water. Petroleum reduces its insulative value by removing natural oils that waterproof the pelage. The rate of heat transfer through fur seal pelts can double after oiling (Kooyman *et al.,* 1976, 1977), adding an energetic burden to the animal. In fact, fouling of approximately one-third of the body surface resulted in 50% greater heat loss in fur seals immersed in water at various temperatures (Kooyman *et al.,* 1976).

Pinnipeds other than fur seals are less threatened by thermal effects of fouling, if at all. Oil has no effect on the relatively poor insulative capacity of sea lion and bearded and ringed seal pelts; oiled Weddell seal samples show some increase in conductance (Oritsland, 1975; Kooyman *et al.,* 1976, 1977). In oil-fouled ringed seals and weaned harp seal pups, core body temperature remained within the normal range, though it was not determined whether there was a compensatory increase in the animals' metabolic rate (Smith and Geraci, 1975; Geraci and Smith, 1976a). If we presume that blubber provides sufficient insulation in these species, we need only be concerned about the thermal effects of oil fouling of newborn phocids, which have little subcutaneous fat and are thought to rely on their lanugo, or birth coat, for insulation. Yet these animals are metabolically equipped to survive birth under rigorous Arctic or sub-Arctic conditions. By utilizing brown fat stores, newborn phocid seals can compensate for the relative ineffectiveness of lanugo made wet by amniotic fluid (Blix *et al.*, 1979). It would be important to know whether these adaptations might offset the potentially deleterious effects of oil fouling until the pups establish adequate blubber.

A coating of oil may have other more obvious effects. Fouling can interfere with locomotion, particularly in young animals. Davis and Anderson (1976) observed two gray seal pups drowning, their "flippers stuck to the sides of their bodies such that they were unable to swim." Similar observations were made following a spill of heavy Bunker C oil in the Gulf of St. Lawrence, where the effect was compounded by low ambient temperatures. A larger stronger seal should be able to resist drowning, but may endure other effects of increased buoyancy due to a tarry coat. Oil might also impair the movements of more delicate structures such as eyelids and vibrissae.

Oil Ingestion

Petroleum is composed of a wide variety of hydrocarbons, some of which are toxic if ingested (Neff, this volume, Chapter 1). Aromatics and other low molecular weight fractions can be absorbed from the intestine and transported via the bloodstream to various target organs. Depending on the amount and composition of the ingested oil, the effects can range from acute death to subtle, progressive organ damage. Ingested hydrocarbons can irritate or destroy epithelial cells that line the stomach and intestine, thereby affecting motility, digestion, and absorption. The effect might be compounded by pre-existing lesions or ulcerations, which frequently occur in pinnipeds infected with gastric nematodes (Geraci and St. Aubin, 1987). Disruptions in the integrity of the mucosa could also facilitate the direct movement of petroleum fractions into the bloodstream. Effects on pinnipeds would presumably parallel those in other mammals.

Ingestion of petroleum hydrocarbons has been implicated in the deaths of a number of stranded gray and harbor seals along the coast of France (Duguy and Babin, 1975; Babin and Duguy, 1985). On postmortem examination, oil was grossly evident within discolored loops of intestine, and its presence was confirmed by gas chromatography in three of seven specimens analyzed. Histopathological changes in liver and intestine were seemingly correlated with the presence of oil in the gut. Yet the authors admitted that carcasses were autolyzed, hence their diagnosis might have been obscured.

The effects of crude oil ingestion have been examined in three studies on two species of phocid seals (Smith and Geraci, 1975; Geraci and Smith, 1976a; Engelhardt, 1982). In all three experiments, relatively small doses of oil were used to duplicate conditions that might realistically occur in the wild; no attempt was made to establish lethal thresholds. Sensitive biochemical tests were used to monitor organ function and detect subtle changes in metabolic processes.

No overtly deleterious effect was noted in harp seal pups given a single dose of up to 75 mL (1–3 mL/kg) of crude oil or in ringed seals given 5 mL of crude oil daily for up to 5 days. Harp seal pups ingesting oil vocalized more than control seals and remained active for several hours after control pups had fallen asleep. Within 1.5 hours, oil was apparent in their feces, suggesting increased gastrointestinal motility. Analysis of tissue-specific enzyme activity in blood revealed mild liver damage in one of six seals given 75 mL of crude oil. The seals were killed and examined at scheduled intervals over a 10-day period following ingestion; no relevant lesions were noted.

Two similar studies on ringed seals focused on absorption, tissue distribution, and clearance of petroleum hydrocarbons. Isotope-labelled benzene (Engelhardt *et al.*, 1977) or naphthalene (Engelhardt, 1982) was added to the oil, and tissues and body fluids were analyzed for levels of radioactivity. Labelled fractions were readily absorbed into the bloodstream, and were detected in liver, blubber, and muscle biopsies first collected 2 days after the initial dose of oil. Hepatic and renal enzyme systems were presumably responsible for the conversion of the labelled fractions to polar metabolites detectable in plasma and urine (Engelhardt, 1982). In the liver of one of four seals tested, activity of aryl hydrocarbon hydroxylase (AHH) was apparently induced to levels four to five times higher than in the other seals; AHH activity was elevated in the kidneys of the other three. Radioactivity in tissues declined rapidly within 2 weeks (Engelhardt *et al.*, 1977), but was still present in trace amounts after 4 weeks.

Plasma activity of tissue-specific enzymes in the seals remained within normal ranges throughout the 10-day monitoring period after oil ingestion (Smith and Geraci, 1975). After four daily doses of crude oil, plasma cortisol levels were markedly increased (Engelhardt, 1982), though without adequate controls it is not possible to ascertain whether this was a consequence of oil ingestion or the stress of repeated handling as the oil was administered. Similarly, the apparent increase in cortisol turnover evident in two of three seals cannot be interpreted as a direct consequence of oil ingestion, since there were no controls for the superimposed stresses of handling.

These limited studies demonstrated that phocid seals can tolerate small quantities of ingested oil. To predict the amount of petroleum which might potentially be toxic, we must extrapolate from data derived for terrestrial species. In rats, the LD_{50} for ingested fuel oils ranges between 5 and 25 mL/kg (Elars Bioresearch Laboratories, Inc., 1980a–d). Thus, a small phocid such as a ringed or harbor seal weighing 50 kg might have to ingest several hundred milliliters of fuel oil to be at risk; for an adult male elephant seal, the quantity would be 10 L or more. It is unrealistic to assume that pinnipeds would consume such large volumes of oil during the course of normal feeding. Nor would grooming present a potential route for ingestion; this activity is relatively uncommon in pinnipeds (McLaren, this volume, Chapter 3). Davis and Anderson (1976) found no evidence of ingested oil in the stomachs of heavily oiled gray seal pups, and no hydrocarbon residues were detected in blood and tissues collected from seals and sea lions at the time of the Santa Barbara spill (Simpson and Gilmartin, 1970). Viscous oil and tar have been noted in the mouths of seals (Anonymous, 1970a) and sea lions (Calkins, 1979), though in these instances the effect would more likely be mechanical interference with feeding than metabolic toxicity.

Chronic ingestion of subtoxic quantities of petroleum may have subtle effects which would only become apparent through long-term monitoring. All pinnipeds examined to date have the enzyme systems necessary to convert absorbed hydrocarbons into polar metabolites which can be excreted in urine (Engelhardt, 1982; Addison and Brodie, 1984; Addison *et al.,* 1986). However, some proportion of the nonpolar fractions will be deposited in lipid-rich tissues, particularly blubber. Petroleum residues have been noted in several species of pinnipeds (Risebrough *et al.,* 1978; Geraci and St. Aubin, 1985), though there is no direct evidence of associated pathologic or metabolic effects. Nor is there evidence that such compounds will accumulate with repeated exposure. Other pollutants that do, including PCBs and DDT, have been implicated in reproductive disorders affecting ringed seals in the Baltic Sea (Helle *et al.,* 1976) and California sea lions (DeLong *et al.,* 1973; Gilmartin *et al.,* 1976). Petroleum hydrocarbon residues might enhance this effect. Mobilization of fat stores during annual molting and reproductive periods could lead to release of residues, and possibly enhance toxicity at those times. Transfer of petroleum fractions via lipid-rich milk is also a potential route of exposure to pups, which have significantly lower levels of some of the detoxifying enzymes (Addison *et al.,* 1986).

Inhalation

There has been no study to assess the effects of inhaled hydrocarbon vapors in pinnipeds. However, indirect evidence from immersion studies and data extrapolated from terrestrial mammals can be used to predict possible consequences in these species. We begin with the basic assumption, for which evidence is accumulating, that pinnipeds have metabolic systems similar to those in other mammals for detoxifying absorbed hydrocarbons (Addison *et al.*, 1986).

Ringed seals placed in a pen containing oil-covered water for 24 hours had levels of petroleum hydrocarbons up to several ppm in blood and tissues (Engelhardt *et al.*, 1977). Since there was no postmortem evidence of ingested oil, the investigators concluded that uptake of hydrocarbons had occurred across the respiratory epithelium. One of the six seals had histological evidence of renal tubular necrosis and fatty degeneration in the liver; a second seal had kidney lesions only (Smith and Geraci, 1975). The occurrence of these lesions correlated with tissue concentrations of petroleum hydrocarbons, and the nature of the damage was similar to that observed in laboratory species (Nau *et al.*, 1966). Plasma levels of a liver-specific enzyme, alanine aminotransferase, were mildly elevated in three of the seals, including the one with morphologic evidence of liver damage. There was no consistent pattern in any of the other plasma chemical constituents analyzed.

Vapor concentrations in the experimental setting were not measured, but likely exceeded those that might occur at sea. Exposure conditions were intensified by the design of the pen, which had plywood walls extending 60–70 cm above water, thereby retarding the dissipation of volatile fractions. Yet the absence of pathologic changes in the lungs of the seals would suggest that the levels were less than 500 ppm. Twenty-four-hour exposure to more concentrated vapors generally results in pulmonary hemorrhage, inflammation, and congestion in a variety of laboratory species (Carpenter *et al.*, 1975, 1976).

In a subsequent immersion study, three ringed seals acclimated to captivity for 2 months died within 71 minutes after oil was introduced to their pool (Geraci and Smith, 1976a). Exposure to gasoline vapor concentrations in excess of 10,000 ppm is rapidly fatal in humans (Machle, 1941), yet it was highly unlikely that such levels could have been attained in the open setting of this experiment. Instead, the combined stress associated with transportation, captivity, and the experimental regime sensitized the seals to a much lower critical threshold. Similar occurrences have been noted occasionally in humans who have been chased after intentionally sniffing gasoline (Bass, 1986). The proximate cause of death in these cases is likely cardiac fibrillation triggered by the synergistic effects of high circulating levels of epinephrine and hydrocarbons.

This observation has significant implications for free-ranging pinnipeds stressed by parasitism or other pre-existing metabolic disorders. In such animals, brief exposure to relatively low concentrations of hydrocarbon vapors might be fatal if combined with other stimuli eliciting a major adrenal response. Parasitic lung disease, a relatively common finding in pinnipeds (Geraci and St. Aubin, 1987), would further complicate the effects of even mild irritation of respiratory tissues. For most pinnipeds, particularly in northern habitats, it is unlikely that petroleum vapors could become sufficiently concentrated to represent a threat. However, selected individuals within a given population may be particularly sensitive and thus be predisposed to the deleterious effects of inhaled hydrocarbon vapors.

Summary

Oil fouling has been implicated in the deaths of pinnipeds, though much of the evidence has been circumstantial. Large-scale mortality has never been observed, even after some of the more catastrophic spills. In general, the prediction that spilled oil would have its greatest impact on young pinnipeds in cold, ice-bound waters has been borne out following the discharge of residual oil in the Gulf of St. Lawrence in 1969.

Pinnipeds show little behavioral or physiologic reactions to the noxious characteristics of oil. Incidental ingestion during feeding, exposure to vapor concentrations that might be expected under natural conditions at sea, and limited surface fouling with relatively fresh oil do not appear to cause significant distress. Pinnipeds trapped near the source of a spill, or forced to emerge in heavy accumulations of oil in leads and around rookeries, exhibit the most severe effects. For fur seals, experimental studies indicate that surface fouling will decrease the insulative value of the pelt, possibly leading to thermal and energetic stress. Individuals of all species and groups that are compromised by pre-existing disease, or stressed by an unfavorable habitat, intra-specific competition, or unusual environmental conditions may be the most vulnerable to the effects of oil exposure.

References

Addison, R. F., and Brodie, P. F. (1984). Characterization of ethoxyresorufin O-de-ethylase in gray seal *Halichoerus grypus*. *Comp. Biochem. Physiol.* **79C,** 261–263.

Addison, R. F., Brodie, P. F., Edwards, A., and Sadler, M. C. (1986). Mixed function oxidase activity in the harbour seal (*Phoca vitulina*) from Sable Is., N.S. *Comp. Biochem. Physiol.* **85C(1),** 121–124.

Anderson, S. S. (1981). Seals in Shetland waters. *Proc. R. Soc. Edinburgh. Sect. B: Biol. Sci.* **80,** 181–188.

Anonymous (1969a). The dead channel. *Time,* Feb. 21, p. 23.

Anonymous (1969b). Seal deaths on isle linked to oil slick. *San Diego Union,* April 5, p. A-7.

Anonymous (1970a). Report of the Task Force—Operation Oil—to the Minister of Transport. *Inf. Can.* **2,** 46–47.

Anonymous (1970b). Nature in Wales. *Mammals* **12**(2):110.

Anonymous (1971a). Arrow oil spill. *Smithson. Inst., Cent. Short-Lived Phenom. Annu. Rep., 1970* Event No. 15-70, pp. 134–136.

Anonymous (1971b). Alaska Peninsula oil spill. *Smithson. Inst., Cent. Short-Lived Phenom. Annu. Rep., 1970* Event No. 36-70, pp. 154–157.

Anonymous (1984). "Appendices to the Log of the Uniaeke G-72 Blowout." Can. Environ. Prot. Serv., Environ. Emergencies Div., Mar. Environ. Br., Atl. Div.

Anonymous (1989). Oil slick could scar Antarctic "for a century." *New Sci.* Feb. 11, p. 31.

Babin, P., and Duguy, R (1985). "Intoxication dûe aux hydrocarbures ingérés par *Halichoerus grypus* et *Phoca vitulina.*" CM1985/N: 12. Cons. Int. Explor. Mer, London.

Baker, J. M., Campodorico, I., Guzman, L., Texera, J. J., and Venegas, C. (1976). An oil spill in the Straits of Magellan. *In* "Marine Ecology and Oil Pollution." (J. M. Baker, ed.), pp. 441–471. Wiley, New York.

Bass, M. (1986). Sniffing gasoline. *J. Am. Med. Assoc.* **255**(19), 2604–2605.

Battelle Memorial Institute (1969). "Review of Santa Barbara Channel Oil Pollution Incident," Rep., Contract No. 14-12-530. Dept. of the Interior and Dept. of Transportation. Richland, Washington.

Blix, A. S., Grav, H. J., and Ronald, K. (1979). Some aspects of temperature regulation in newborn harp seal pups. *Am. J. Physiol.* **236**(3), R188-R197.

Bonner, W. N., and Hickling, G. (1971). The grey seals of the Farne Islands. *Nat. Hist. Soc. Tyne Northumberl.,* Durham Newcastle Upon Trans. Tyne **17,** 141–162.

Bourne, W. R. P. (1979). The *Christos Bitas* affair. *Mar. Pollut. Bull.* **10**(5), 122–123.

Brownell, R. L., Jr., and Le Boeuf, B. J. (1971). California sea lion mortality: natural or artifact? *Biol. and Bacteriol.* **1,** 287–306.

Calkins, D. G. (1979). "Marine Mammals of Lower Cook Inlet and the Potential for Impact from Outer Continental Shelf Oil and Gas Exploration, Development and Transport," Final Report. Outer Continental Shelf Environ. Assess. Program, pp. 171–264. U.S. Dept. of the Interior, Minerals Manage. Serv., Anchorage, Alaska.

Carpenter, C. P., Kinkead, E. R., Geary, D. L., Sullivan, L. J., and King, J. M. (1975). Petroleum hydrocarbon toxicity studies. V. Animal and human responses to vapors of mixed xylenes. *Toxicol. Appl. Pharmacol.* **33,** 543–558.

Carpenter, C. P., Kinkead, E. R., Geary, D. L., Sullivan, L. J., and King, J. M. (1976). Petroleum hydrocarbon toxicity studies. XIII. Animal and human response to vapors of toluene concentrate. *Toxicol. Appl. Pharmacol.* **36,** 473–490.

Cowles, C. J., Hansen, D. J., and Hubbard, J. D. (1981). "Types of Potential Effects of Offshore Oil and Gas Development on Marine Mammals and Endangered Species of the Northern Bering Sea and Arctic Ocean," Tech. Pap. No. 9. Bur. Land Manage., Alaska Outer Continental Shelf Region, Anchorage.

Davies, J. L. (1949). Observations on the gray seal (*Halichoerus grypus*) at Ramsay Island, Pembrokeshire. *Proc. Zool. Soc. London* **119,** 673–692.

Davis, A., Schafer, L. J., and Bell, Z. G. (1960). The effects on human volunteers of exposure to air containing gasoline vapors. *Arch. Environ. Health* **1,** 548–554.

Davis, J. E., and Anderson, S. S. (1976). Effects of oil pollution on breeding gray seals. *Mar. Pollut. Bull.* **7,** 115–118.

DeLong, R. L., Gilmartin, W. G., and Simpson, J. G. (1973). Premature births in California sea lions: Association with high organochlorine pollutant residue levels. *Science* **181,** 1168–1170.

Duguy, R., and Babin, P. (1975). "Acute Intoxication by Hydrocarbons Observed in a Harbor Seal (*Phoca vitulina*)," Rep. C.M. 1975/N : 5. Cons. Int. Explor. Mer, Comité de Mammifères Marins.

Easton, R. (1972). "Black Tide: The Santa Barbara Oil Spill and its Consequences." Delcorte Press, New York.

Elars Bioresearch Laboratories, Inc. (1980a). "Acute Toxicity Tests of API 78-2 No. 2 Home Heating Oil (30% Cat)," Proj. 1443, Med. Res. Publ. 2732771. Am. Pet. Inst., New York.

Elars Bioresearch Laboratories, Inc. (1980b). "Acute Toxicity Tests of API 78-7 No. 6 Heavy Fuel Oil (API Gravity 17.1/0.8% S)," Proj. 1443, Med. Res. Publ. 2732774. Am. Pet. Inst., New York.

Elars Bioresearch Laboratories, Inc. (1980c). "Acute Toxicity Tests of API 79-2 No. 6 Heavy Fuel Oil (API Gravity 5.2/1.2% S)," Proj. 1443, Med. Res. Publ. 2732813. Am. Pet. Inst., New York.

Elars Bioresearch Laboratories, Inc. (1980d). "Acute Toxicity Tests of API 78-6 No. 6 Heavy Fuel Oil (API Gravity 11.7/2.7% S)," Proj. 1443, Med. Res. Publ. 2732814. Am. Pet. Inst., New York.

Engelhardt, F. R. (1982). Hydrocarbon metabolism and cortisol balance in oil-exposed ringed seals, *Phoca hispida*. *Comp. Biochem. Physiol.* **72C,** 133–136.

Engelhardt, F. R., Geraci, J. R., and Smith, T. J. (1977). Uptake and clearance of petroleum hydrocarbons in the ringed seal, *Phoca hispida*. *J. Fish. Res. Board Can.* **34,** 1143–1147.

Environmental Sciences Ltd. (ESL) (1981). "A Prospectus on the Biological Effects of Oil Spills in Marine Environments," Doc. No. BEISSD 13. Dome Petroleum Ltd., Calgary.

Fogden, S. C. L. (1971). Mother-young behavior at gray seal breeding beaches. *J. Zool.* **164,** 61–92.

Geraci, J. R., and St. Aubin D. J. (1985). "Study of the Effects of Oil on Cetaceans," Final Rep., Contract No. AA551-CT9-29. U.S. Dept. of the Interior, Bureau of Land Manage., Washington, D.C.

Geraci, J. R., and St. Aubin, D. J. (1987). Effects of parasites on marine mammals. *Int. J. Parasitol.* **17**(2), 407–414.

Geraci, J. R., and Smith, T. G. (1976a). Direct and indirect effects of oil on ringed seals (*Phoca hispida*) of the Beaufort Sea. *J. Fish. Res. Board Can.* **33**(9), 1976–1984.

Geraci, J. R., and Smith, T. G. (1976b). Behavior and pathophysiology of seals exposed to crude oil. *In* "Sources, Effects and Sinks of Hydrocarbons in the Aquatic Environment," pp. 448–462. Am. Inst. Biol. Sci., Washington, D.C.

Gill, C., Booker, F., and Soper, T. (1967). "The Wreck of the *Torrey Canyon*." David & Charles Publ., Newton Abbot, U.K.

Gilmartin, W.G., DeLong, R. L., Smith, A. W., Sweeney, J. C., DeLappe, B. W., Risebrough, R. W., Griner, L. A., Dailey, M. D., and Peakall, D. B. (1976). Premature parturition in the California sea lion. *J. Wildl. Dis.* **12,** 104–115.

Grose, P. L., Mattson, J. S., and Petersen, H. (1979). Marine mammals and sea birds. *U.S. NTIS, PB Rep.* **PB80-173727,** 9-1–9-8.

Hansbrough, J. F., Zapata-Sirvent, R., Dominic, W., Sullivan, J., Boswick, J., and Wang, X.-W. (1985). Hydrocarbon contact injuries. *J. Trauma* **25,** 250–252.

Helle, E., Olsson, M., and Jensen, S. (1976). PCB levels correlated with pathological changes in seal uteri. *Ambio* **5,** 261–263.

Hess, R., and Trobaugh, L. (1971). Kodiak Island oil pollution. *Smithson. Inst., Cent. Short-Lived Phenom., Annu. Rep., 1970* Event 26-70, pp. 150–153.

Jamieson, G. S., and Fisher, H. D. (1972). The pinniped eye: A review. *In* "Functional Anatomy of Marine Mammals." (R. J. Harrison, ed.), Vol. 1, pp. 245–261. Academic Press, London.

King, J. E. (1983). "Seals of the World," 2 ed. Cornell Univ. Press, Ithaca, New York.

Klauder, J. V., and Brille, F. A. (1947). Correlation of boiling ranges of some petroleum solvents with irritant action on the skin. *Arch. Dermatol. Syph.* **56,** 197–215.

Kooyman, G. L., Gentry, R. L., and McAllister, W. B. (1976). "Physiological Impact of Oil on Pinnipeds," Rep. N.W. Fish. Cent., Natl. Mar. Fish. Serv., Seattle, Washington.

Kooyman, G. L., Davis, R. W., and Castellini, M. A. (1977). Thermal conductance of immersed pinniped and sea otter pelts before and after oiling with Prudhoe Bay crude. *In* "Fate and Effects of Petroleum Hydrocarbons in Marine Ecosystems and Organisms" (D. A. Wolfe, ed.), pp. 151–157. Pergamon, New York.

Le Boeuf, B. J. (1971). Oil contamination and elephant seal mortality: A "negative" finding. *Biol. Bacteriol.* **1,** 277–285

Lillie, H. (1954). Comments in discussion. *Proc. Int. Conf. Oil Pollut.,* London, **1953,** pp. 31–33.

Lupulescu, A. P., Birmingham, D. J., and Pinkus, H. (1973). An electron microscopic study of human epidermis after acetone and kerosene administration. *J. Invest. Dermatol.* **60,** 33–45.

Machle, W. (1941). Gasoline intoxication. *J. Am. Med. Assoc.* **117,** 1965–1971.

Mansfield, A. W. (1970). "Field Report of Seal Investigations in Chedabucto Bay and at Sable Island, Nova Scotia 2 March-7 April 1970." Fish. Res. Board Can., Arct. Biol. Stn., Ste. Anne de Bellevue, P.Q.

McMillan, I. I. (1969). Another look at the big slick. *Defenders Wildl. News* **44,** 149–153.

Muller-Willie, L. (1974). How effective is oil pollution legislation in arctic waters? *Musk-Ox* **14,** 56–57.

Nachtigall, P. E. (1986). Vision, audition, and chemoreception in dolphins and other marine mammals. *In* "Dolphin Cognition and Behavior: A Comparative Approach." (R. J. Schusterman, J. A. Thomas, and F.G. Wood, eds.), pp. 79–113. Erlbaum, Hillsdale, New Jersey.

Nau, C. A., Neal, J., and Thornton, M. (1966). C9-C12 fractions obtained from petroleum distillates. An evaluation of their potential toxicity. *Arch. Environ. Health* **12,** 382–393.

Odell, D. K. (1970). Premature pupping in the California sea lion *Proc. 7th Annu. Biosonar Diving Mammal Conf.,* Stanford Res. Inst., pp. 185–190. Menlo Park, California.

Øritsland, N. A. (1975). Insulation in marine mammals: The effect of crude oil on ringed seal pelts. *In* "The Effect of Contact and Ingestion of Crude Oil on Ringed Seals of the Beaufort Sea." (T. G. Smith and J.R . Geraci, eds.), Beaufort Sea Proj. Tech. Rep. No. 5, pp. 48–67. Inst. Ocean Sci., Sidney, British Columbia.

Parsons, J., Spry, J., and Austin, T. (1980). Preliminary observations on the effect of Bunker C fuel oil on seals on the Scotian shelf. *Proc. Workshop, Bedford Inst. Oceanogr.,* Dartmouth, Nova Scotia, pp. 193–202.

Prieur, D., and Hussenot, E. (1978). Marine mammals stranded during the *Amoco Cadiz* oil spill. *Extrait de Penn ar Bed* **11**(94), 361–364.

Reiter, G.A. (1981). Cold weather response F/V *Ryuyo Maru* No. 2, St. Paul, Pribiloff Islands, Alaska. *In* "Proceedings of the 1981 Oil Spill Conference," Publ. No. 4334, pp. 227–231. Am. Petrol Inst., Washington, D.C.

Renouf, D., and Davis, M. B. (1982). Evidence that seals may use echolocation. Nature (London) **300,** 635–637.

Renouf, D., Galway, G., and Gaborko L, (1980). Evidence of echolocation in harbour seals. *J. Mar. Biol. Assoc. U.K.* **60,** 1039–1042.

Renouf, D., Lawson, J., and Gaborko, L. (1983). Attachment between harbor seal (*P. vitulina*) mothers and pups. *J. Zool.* **199,** 179–187.

Risebrough, R. W., Walker, W., II, Springer, A.M., Clayton, J. R., Letterman, E.F., Payne, J. R., and Schmidt, T. T. (1978). "A Search for Pollutants of Petroleum Origin in Tissues of Harbor Seals, *Phoca vitulina*, in San Francisco Bay" Final Rep.,Contract No. MM-7AC007. Mar. Mammal Comm., Washington, D.C.

Sandegren, F. E. (1970). Breeding and maternal behavior of the Steller sea lion (*Eumetopias jubata*) in Alaska. M.Sc. Thesis, University of Alaska, Anchorage, AK.

Sergeant, D. E. (1987). "Harp Seals, Man and Ice." Canadian Department of Fisheries and Oceans, Arctic Biological Station. (unpublished manuscript).

Shaughnessy, P. D., and Chapman, P. (1984). Commensal Cape fur seals in Cape Town docks. *S. Afr. J. Mar. Sci.* **2,** 81–91.

Simpson, J. G. (1970). Elephant seals and California sea lions apparently not killed by Santa Barbara oil spill. *Biol. Conserv.* **2**(2), 89.

Simpson, J. G., and Gilmartin, W. G. (1970). An investigation of elephant seal and sea lion mortality on San Miguel Island. *BioScience* **20**(5), 289.

Smith, T. G., and Geraci, J. R. (1975). "The Effect of Contact and Ingestion of Crude Oil on Ringed Seals of the Beaufort Sea," Beaufort Sea Proj., Tech. Rep. No. 5. Inst. Ocean Sci., Sidney, British Columbia.

Snell, D. (1969). Iridescent gift of death. *Life* **66**(23), 22–27.

Spooner, M. F. (1967). Biological effects of the *Torrey Canyon* disaster. *J. Devon Trust Nat. Conserv.*, pp. 12–19.

Tagami, H., and Ogino, A. (1973). Kerosene dermatitis; factors affecting skin irritability to kerosene. *Dermatologica* **146,** 123–131.

van Haaften, J. L. (1973). Die Bewirtschaftung von Seehunden in den Neiderlanden [Seal management in the Netherlands]. *Beitr. Jagd-Wildforschung* **8,** 345–349.

Walsh, W. A., Scarpa, F. J., Brown, R. S., Ashcraft, K. W., Green, V. A., Holder, T. M., and Amoury, R. A. (1974). Gasoline immersion burn. *N. Engl. J. Med.* **291,** 830.

Warner, R. E. (1969). "Environmental Effects of Oil Pollution in Canada: An Evaluation of Problems and Research Needs." Rep. Can. Wildl. Serv., Ottawa.

Wartzok, D., Schusterman, R., and Gailey-Phipps, J. (1984). Seal echolocation. *Nature (London)* **308,** 753.

5

Cetaceans and Oil: Ecologic Perspectives

Bernd Würsig*
Moss Landing Marine Laboratories
Moss Landing, California 95039

Introduction

Cetaceans are derived from terrestrial protoungulates, and it is thought that they evolved from early carnivorous ungulates called mesonychids over 50 million years ago (Barnes, 1984). Modern pigs are rather closely related to cetaceans. Imagine the evolutionary forces that changed four-legged terrestrial predators into sleek, hairless, legless mammals of the sea. As a group, cetaceans have invaded water more extensively than have other marine mammals. They now occur in all oceans, from the tropics to the ice edges, and one family is found in some major river systems as well. Only pelagic depths are unavailable to them, although the sperm whale is known to dive to depths of over one kilometer (Slijper, 1979).

The order Cetacea is divided into two suborders: the mysticetes or baleen whales, and the odontocetes or toothed whales, dolphins, and porpoises. These two major forms appear to have separated approximately 30 million years ago from the now extinct archaeocetes. The baleen whales comprise 11 species in three families: the Balaenidae or right and bowhead whales (four species), the Balaenopteridae or rorqual whales (six species), and the Eschrichtiidae or gray whale. All baleen whales are large. They range in size from the little-known pygmy right whale, about 5 m long, to the blue whale, from 25 to 30 m long—the largest living animal.

The right and bowhead whales are chunky animals, not particularly built for speed. They have broad backs and large powerful tails. Their

*Present address: Department of Marine Biology, Texas A & M University, Galveston, Texas 77553

Sea Mammals and Oil: Confronting the Risks
Copyright © 1990 by Academic Press, Inc.
All rights of reproduction in any form reserved.

most remarkable feature is a huge head, which in adult bowhead whales may be almost one-third the length of the body. On the other hand, rorqual whales, represented by blue, fin, sei, humpback, minke, and Bryde's whales, are thin and sleek, with an almost perfectly streamlined shape. Gray whales have an intermediate form, neither fat nor sleek.

The toothed whales, dolphins, and porpoises comprise a large and diverse taxonomic group, from the small river dolphins and coastal porpoises to the sperm whale, the largest toothed animal. Although taxonomic schemes vary, most identify 68 species in six families. The largest group (34 species) is the Delphinidae, which includes five subfamilies of coastal and oceanic dolphins. They range in size from the miniature tucuxi to the formidable killer whale. The other families are the Physeteridae or sperm whales (three species), Monodontidae or narwhal and beluga (two species), Ziphiidae or beaked whales (18 species), Phocoenidae or true porpoises (six species), and Platanistidae or river dolphins (five species) (Rice, 1977). Most toothed whales are gregarious, usually occurring in groups numbering a few to many dozens of individuals. Pelagic Delphinidae sometimes occur in groups of several thousand animals (Nishiwaki, 1975), reminiscent of herds of terrestrial ungulates.

Population estimates for selected cetacean species found in North American waters are given in Table 5-1. For many, the estimates are crude, reflecting the difficulty in censusing dispersed pelagic animals. Technological advances will surely improve our knowledge of cetacean abundance, but the size of a population is not the sole criterion for assessing its status or vulnerability to the effects of oil pollution. Discrete subpopulations may be locally threatened by direct exploitation, habitat deterioration, fishery-related mortality, or increased competition for fisheries' resources. For example, Dall's porpoises number in the millions, yet they are locally threatened by hunting in Japanese waters (Brownell *et al.*, 1989); belugas in the St. Lawrence River are a relict population in need of protection, whereas other subpopulations in the Arctic are abundant.

The International Union for the Conservation of Nature and Natural Resources (IUCN) and the government of the United States, under the Endangered Species Act (ESA), use a number of criteria to identify a species as threatened (ESA), vulnerable (IUCN), or endangered (both organizations), and these designations have been included in Table 5-1. Recently, there has been a call to reexamine the classification of cetacean species, because some listed under the ESA, including the sperm and gray whales, are in no danger, whereas others that are nearly extinct or severely threatened are not included (Brownell *et al.*, 1989). A more rigorous application of the criteria for designating status will be required to guide oil-exploration activities.

Table 5–1
Population Estimates of Cetaceans Found in North American Waters

Species	Geographic Stock[a]	Population	Status[b]	Sources[c]
Family Balaenidae				
Bowhead Whale	W. Arctic	7,800		Zeh *et al.* (1988)
	E. Canadian Arctic	<700		Mitchell and Reeves (1981)
	Worldwide		Endangered	USESL, IUCN
Northern Right Whale	W.N. Atlantic	<350		Mate (1989)
	E.N. Pacific	<several hundred		Leatherwood *et al.* (1982)
	Worldwide	3,100–3,200		NMFS (1988)
			Endangered	USESL, IUCN
Family Balaenopteridae				
Blue Whale	W.N. Atlantic	100		NMFS (1988)
	N. Pacific	1,600		NMFS (1988)
		1,400–1,900		Braham (1984)
	Worldwide	11,700		NMFS (1988)
			Endangered	USESL, IUCN
Bryde's Whale	Worldwide	30,000–56,000		NMFS (1988)
			Not listed	USESL, IUCN
Fin Whale	W.N. Atlantic	3,600–6,300		NMFS (1988)
	Newfoundland	1,900		Meredith and Campbell (1988)
	Nova Scotia	430		Meredith and Campbell (1988)
	Greenland	7,200		Meredith and Campbell (1988)
	E.N. Pacific	15,000–19,000		NMFS (1988)
	Worldwide	105,000–122,000		NMFS (1988)
			Endangered	USESL
			Vulnerable	IUCN

(continued)

Table 5–1 (*Continued*)

Species	Geographic Stock[a]	Population	Status[b]	Sources[c]
Humpback Whale	W.N. Atlantic	5,800		NMFS (1988)
	N. Pacific	1,200–9,500		NMFS (1988)
	Worldwide	9,500–10,000		NMFS (1988)
			Endangered	USESL, IUCN
Minke Whale	Worldwide	315,800–331,800		NMFS (1988)
			Not listed	USESL, IUCN
Sei Whale	W.N. Atlantic	2,600		NMFS (1988)
	N. Pacific	22,000–37,000		MNFS (1988)
	Worldwide	48,000–63,000		NMFS (1988)
			Endangered	USESL
			Not listed	IUCN
Family Eschrichtiidae				
Gray Whale	E.N. Pacific	21,000		Brownell *et al.* (1989)
			Endangered	USESL
			Not listed	IUCN
Family Physeteridae				
Sperm Whale	E.N. Pacific	274,000		MNFS (1988)
	Worldwide	982,000		NMFS (1988)
			Endangered	USESL
			Not listed	IUCN
Family Monodontidae				
Beluga Whale	MacKenzie Delta	7,000–11,500		Braham (1984)
	Hudson Bay	25,000		P. Richard (personal communication)
	Frobisher Bay	250–300		Braham (1984)
	Cumberland Sound	600		Braham (1984)
	Canadian High Arctic	6,300–18,600		Smith *et al.* (1985)

	Alaska:			
	Norton Sound/Yukon Delta	1,000–2,000		K.J. Frost (personal communication)
	Bristol Bay/Kuskokwim Bay	1,000–1,500		Braham (1984)
	Cook Inlet	600–1,000		Braham (1984)
	E. Chukchi Sea	2,500–3,000		K.J. Frost (personal communication)
	St. Lawrence Estuary	400–500		Sergeant (1986)
	Worldwide	62,000–88,000		NMFS (1985)
		50,000		Brownell *et al.* (1989)
		40,000–55,000		
			Not listed	USESL, IUCN
Narwhal	Canadian E. Arctic	29,000		IWC (1986)
	Worldwide	30,000		NMFS (1988)
		35,000		Brownell *et al.* (1989)
			Not listed	USESL, IUCN
Family Delphinidae				
Atlantic White-sided Dolphin	W.N. Atlantic	24,000		NMFS (1988)
			Not listed	USESL, IUCN
Bottlenose Dolphin	W.N. Atlantic	14,000–23,000		NMFS (1988)
	Florida & Gulf of Mexico	10,000		Schmidly (1981)
	Florida:Banana R. & Indian R.	200–600		IWC (1986)
	California: San Diego County	240		IWC (1986)
	Worldwide		Not listed	USESL, IUCN

(continued)

Table 5–1 (*Continued*))

Species	Geographic Stock[a]	Population	Status[b]	Sources[c]
Common Dolphin	W.N. Atlantic	31,000		NMFS (1988)
	E.N. Pacific	900,000		NMFS (1988)
	S. California	15,000–57,000		Dohl *et al.* (1986)
	Worldwide		Not listed	USESL, IUCN
Killer Whale	Puget Sound & Vancouver Is.	330		Bigg *et al.* (1987)
	S. Alaska	600		IWC (1987)
	Worldwide		Not listed	USESL, IUCN
Long-finned Pilot Whale	E. Newfoundland	13,000–50,000		IWC (1987)
	Worldwide		Not listed	USESL, IUCN
Pacific White-sided Dolphin	E.N. Pacific	50,000		NMFS (1988)
	Worldwide		Not listed	USESL, IUCN
Risso's Dolphin	W.N. Atlantic	10,000		NMFS (1988)
	Worldwide		Not listed	USESL, IUCN
Spinner Dolphin	E.N. Pacific	1.1 million		Buckland and Anganuzzi (1988)
	Worldwide		Not listed	USESL, IUCN
Spotted Dolphin	E.N. Pacific	3.3 million		Buckland and Anganuzzi (1988)
	Worldwide		Not listed	USESL, IUCN
Striped Dolphin	E.N. Pacific	2.3 million		NMFS (1988)
	Worldwide		Not listed	USESL, IUCN

Family Phocoenidae				
Dall's Porpoise	E.N. Pacific	607,000 to 1.2 million		Jones *et al.* (1987)
	Bering Sea	215,000–392,000		Jones *et al.* (1987)
	W. Central Pacific	622,000 to 1.2 million		Jones *et al.* (1987)
	Worldwide	1.4–2.8 million		NMFS (1988)
			Not listed	USESL, IUCN
Harbor Porpoise	W.N. Atlantic	18,000		NMFS (1988)
	Worldwide		Not listed	USESL, IUCN
Vaquita	Gulf of California (entire population)	200–300		J. Barlow (personal communication)
			Endangered	USESL

[a] Designation of a geographic stock does not imply complete isolation from other groups in adjoining water.
[b] Status according to the U.S. Endangered Species List (USESL) or International Union for the Conservation of Nature and Natural Resources (IUCN), Gland, Switzerland.
[c] The source populations are generally the latest original references to population size and status. Secondary sources are used where original estimates are qualified or where the primary sources are relatively inaccessible. NMFS, National Marine Fisheries Service.

Distribution

Cetaceans occur worldwide, and some species are truly cosmopolitan. Most notable are the mysticete minke whale and the odontocete killer whale, both of which are widely distributed in polar, temperate, tropical, nearshore, and pelagic waters. The majority, however, are restricted to more localized habitats, giving each of the major bodies of water surrounding North America a distinctive cetacean fauna.

Bowhead whales, prized for their baleen and abundant oil and hunted to near extinction by the 19th century whalers, are the northernmost mysticetes, always living near ice in the Bering, Chukchi, and Beaufort seas of the western Arctic, and in the Baffin Bay, Hudson Bay, and Labrador Sea areas of the eastern Arctic (Fig. 5–1). They have well-defined migration routes, often restricted by extensive ice cover (Reeves *et al.*, 1983; Braham *et al.*, 1984). Oil development activities occur along their main migratory and feeding routes in the western Arctic (Richardson *et al.*, 1987). Narwhals and belugas coexist with bowheads in part of their range. Narwhals, with their characteristic "unicorn" tusk, are mainly restricted to Canadian waters of the eastern Arctic, while strikingly white beluga whales occur on both sides of the continent, as far south as British

A spinner dolphin, named for such aerial antics, breaks the water off Hawaii. Photo by B. Würsig.

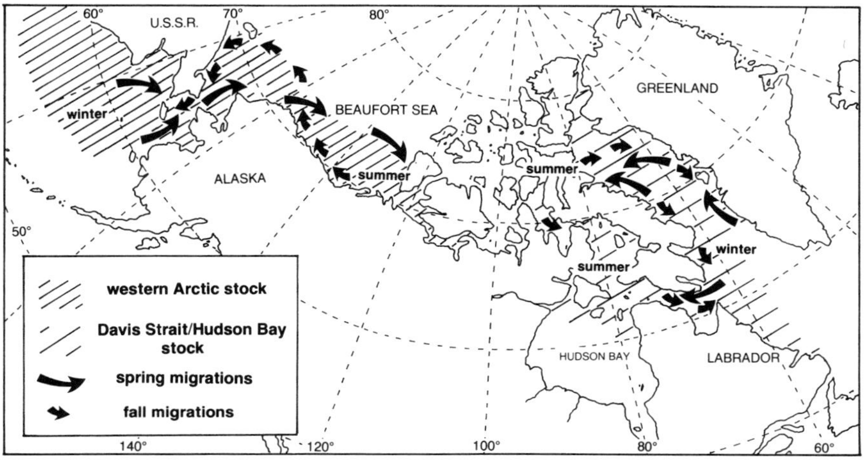

Figure 5-1
Distribution of bowhead whales.

Columbia on the Pacific side and along the Canadian maritime provinces on the Atlantic (Reeves and Katona, 1980), with one discrete stock in the Gulf of St. Lawrence (Sergeant, 1973; Sergeant and Hoek, 1974).

Gray whales occur in the Bering, Chukchi, and occasionally the Beaufort Sea (Rugh and Fraker, 1981; Berzin, 1984) in summer and as far south as midMexico in the Pacific Ocean in winter (Fig. 5–2) (Norris *et al.*, 1983). Blue, fin, and humpback whales may be found anywhere from the lower Bering Sea to the shores of Mexico, but seem to concentrate seasonally in certain areas. For example, blue and humpback whales are seen around the Farallon Islands off central California in late summer (Dohl *et al.*, 1983); humpback whales occur in the Gulf of Alaska and in fjords of the Alaskan panhandle in summer (Fig. 5–3) (Hall, 1979; Rice and Wolman, 1981); fin whales occur in the northern Gulf of California year-round (personal observation). Minke whales are widely distributed in the Bering Sea, Gulf of Alaska, inshore waters of British Columbia and Washington, and along the United States' west coast. They are seldom seen north of the Bering Strait (Rice, 1974). Sei whales occur from California northward to the Gulf of Alaska, but they favor oceanic over coastal waters (Masaki, 1977; Leatherwood *et al.*, 1982).

The Pacific coast of North America has features that influence the pattern of cetacean distibution. The coastline of the North Pacific is generally rocky and abrupt with a narrow or nonexistent continental shelf. This allows for pelagic forms to come relatively close to shore. As well, the cold

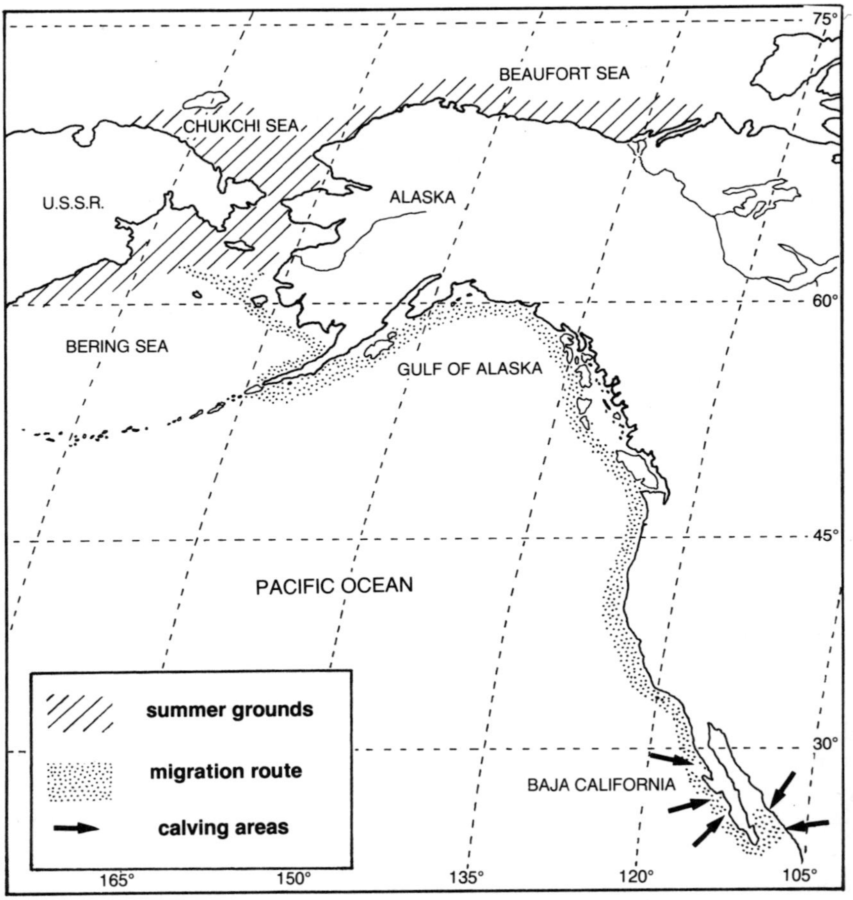

Figure 5-2
Distribution of gray whales.

California current sweeps down from the Alaskan Panhandle, and brings northern forms as far as Baja California, Mexico. Thus, harbor porpoises occur close to shore in bays and estuaries from the Bering Strait south to Point Conception, California (Norris and Prescott, 1961; Gaskin *et al.*, 1974), and Dall's porpoises from the northern Bering Sea to the Mexican border (Morejohn, 1979; Jefferson, 1988). Pacific white-sided dolphins range from the Aleutian chain to the southern tip of the Baja Peninsula (Leatherwood *et al.*, 1984); northern right whale dolphins have a similar distribution, although they do not frequent waters north of British Colum-

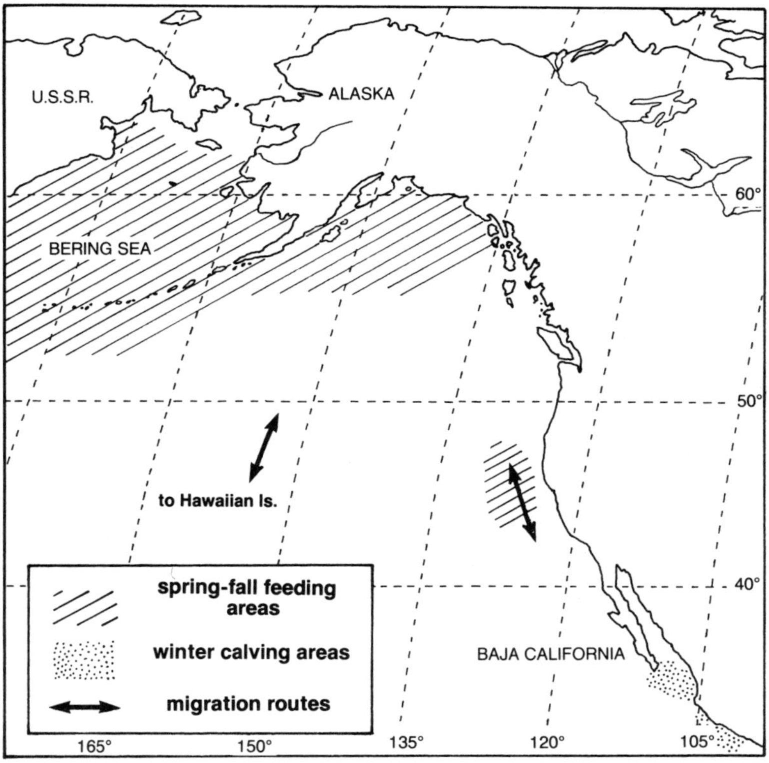

Figure 5-3
Distribution of humpback whales, North Pacific.

bia or south of northern Baja (Leatherwood and Walker, 1979). Dall's porpoises, Pacific white-sided dolphins, and northern right whale dolphins inhabit generally deeper water than harbor porpoises, ranging hundreds of kilometers from the coast, occasionally coming close to shore over submarine canyons and off headlands (Leatherwood *et al.,* 1982).

Common dolphins occur from the tropics to the limits of warm temperate waters in all oceans, as far north as northern California in the summer (Evans, 1982a; Dohl *et al.,* 1986). Risso's dolphins (Leatherwood *et al.,* 1980) and short-finned pilot whales (Norris and Prescott, 1961) have a similar distribution in the Pacific, while bottlenose dolphins rarely venture north of Point Conception, California (Walker, 1981). However, in recent years, bottlenose dolphins have followed abnormal warm water incursions as far north as Monterey Bay, California, where they are now commonly sighted (Wells *et al.,* 1990). Killer whales are uncommon near shore except in waters off Washington State and British Columbia and in the Gulf of Alaska where they feed on abundant stocks of salmon (Matkin and Leatherwood, 1986; Heyning and Dahlheim, 1988). Spinner and spotted dolphins of the genus *Stenella*, possibly the most abundant of the Cetacea, are found offshore in the eastern tropical Pacific.

The Atlantic coast is physically unlike that of the Pacific. The coastline slopes gently and the shelf is wide. Furthermore, the Gulf Stream allows warm temperate species to extend to higher latitudes than in the Pacific. Humpback (Fig. 5–4), blue, fin, sei, and minke whales can be found anywhere from Florida to the Canadian Maritime Provinces and north to Greenland, but tend to concentrate seasonally in certain areas such as the Gulf of St. Lawrence and the banks off New England (Caldwell and Golley, 1965; Sergeant *et al.,* 1970; Sergeant, 1977). Northern right whales, while not abundant (Winn *et al.,* 1981), congregate close to shore off New England in summer and off Georgia and Florida in winter (Kraus *et al.,* 1986).

Short-finned pilot whales range from the tropics to Delaware Bay; long-finned pilot whales prefer cold temperate waters north of Cape Hatteras (Mitchell, 1975; Leatherwood *et al.,* 1976), whereas white-beaked and Atlantic white-sided dolphins are found still further north. The familiar bottlenose dolphins, exhibited worldwide and extensively studied for the past 35 years, occur from the Canadian Maritime Provinces southward along the United States coastline, as two distinguishable subpopulations: an inshore group with a more southerly distribution, and a pelagic stock which ranges further north (Scott *et al.,* 1988). Risso's dolphins share the same distribution as the offshore bottlenose dolphins. Harbor porpoises occupy cold temperate waters from North Carolina to the Arctic, generally close to shore in shallow waters. Common, spotted, and striped dolphins occur throughout the tropical and warm temperate Atlantic, the striped dolphins generally far offshore. Spinner dolphins are mainly tropical and subtropical and hardly ever venture north of Florida. Killer whales occasionally visit the North Atlantic coastline (Reeves and Mitchell, 1987).

The Gulf of Mexico is unlike the other seas bordering the continent. It is a shallow, warm basin having low productivity and a high rate of siltation. It is home to many cetacean species, including the tropical Bryde's whale, short-finned pilot whale, several species of the genus *Stenella*, common dolphin, Risso's dolphin, and sperm whale. The most numerous inshore resident is the Atlantic bottlenose dolphin (Fig. 5–5) (Shane, 1980; Wells *et al.,* 1980; Shane *et al.,* 1986). Humpback whales occur in the Gulf's eastern edge during their winter breeding season. Fin

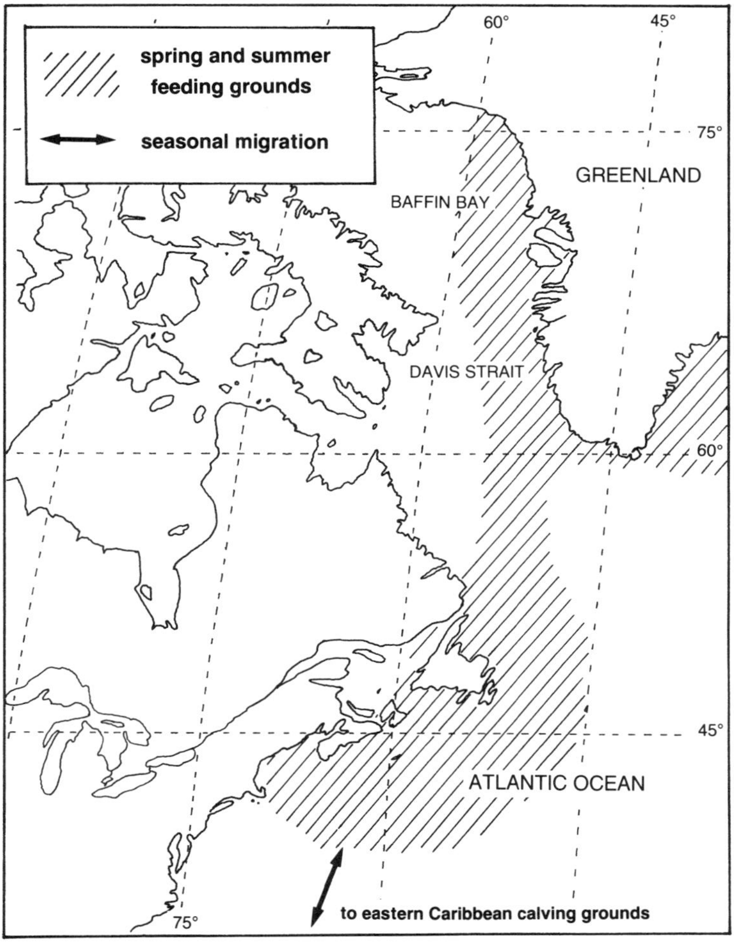

Figure 5-4
Distribution of humpback whales, western North Atlantic.

and right whales are there in winter as well, but are generally farther from shore.

Several other species, including sperm whales, beaked whales, and pelagic dolphins tend to stay in deep, usually pelagic, waters off North America. Although these species may encounter oil (especially as oil exploration moves into deeper water on the continental slope and rise), the chances for encounters are considerably less than for inshore forms.

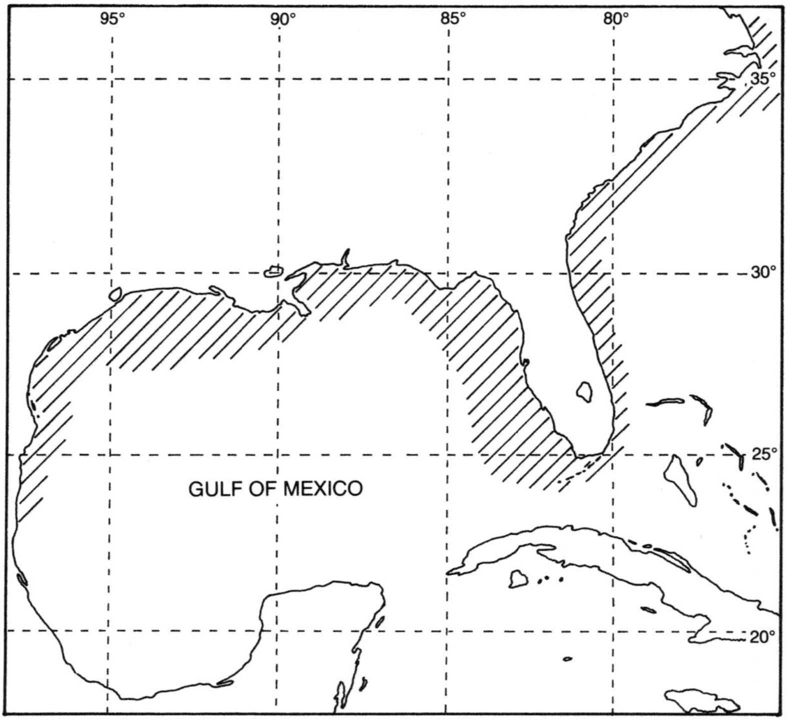

Figure 5-5
Inshore distribution of Atlantic bottlenose dolphins in the Gulf of Mexico and south Atlantic regions. Bottlenose dolphins probably occur on all shores marked by this map, but information is sparse for all but the lined area.

Life Histories

Reproduction, Growth and Maturation

Cetaceans give birth to single young, after gestation periods varying from about 11 months in the small harbor porpoise (Gaskin *et al.*, 1984) to 15 months in the much larger pilot, killer, and sperm whales (Best *et al.*, 1984; Kasuya and Marsh, 1984). Most of the medium-sized odontocetes, such as the bottlenose dolphin, have a gestation time of approximately 11–12 months (Sergeant *et al.*, 1973; Perrin and Reilly, 1984). Duration of pregnancy tends to increase with size in the toothed whales, a correlation that does not hold for baleen whales, these being constrained to gestation periods of about 1 year. Seasonal migrations take them to cold-water feeding grounds in the summer, leaving only a narrow seasonal window for reproduction.

This migratory cycle of baleen whales also places constraints on duration of lactation, which in baleen whales is completed in less than a year (Tyack, 1986). In odontocetes, lactation ranges from about 7 months in the harbor porpoise to well over 1 year in most others. Short gestation and lactation periods in baleen whales are part of a tremendously rapid growth which continues to adulthood. The large blue whale, for example, has a gestation period of 11 months, is weaned at 7 months, and is sexually mature at 5 years (Lockyer, 1984). Sperm whales, on the other hand, are born after 15 months' gestation, are weaned at approximately 2 years (some few suckling as long as 13 years), and take from 9 years (females) to 20 years (males) to become sexually mature (Best *et al.*, 1984). There is some evidence that baleen whales whose populations have been reduced by human exploitation showed a decline of 3–6 years in the age of sexual maturity. The data are most convincing for several populations of fin (Ohsumi, 1986), sei, and minke whales (Brown and Lockyer, 1984). Nevertheless, size at which these whales reach sexual maturity has remained constant, indicating that they are growing faster, possibly because of greater food reserves available, especially in the Antarctic (Lockyer, 1981).

Most toothed whales larger than dolphins and porpoises have intercalf intervals greater than 3 years (with 6 years the norm for sperm whales), but most baleen whales give birth every 2 years. Bowhead and right whales are an exception; they have intercalf intervals generally 3 years or longer (Payne and Dorsey, 1983; Nerini *et al.*, 1984). Rapid development and high calf production in most baleen whales suggest they are relatively short-lived, but aging techniques are not sufficiently reliable to test the hypothesis. Slower developing odontocetes, such as sperm, killer, and pilot whales, can reach ages of 50 years, and possibly more (Best, 1979; Perrin and Reilly, 1984). Longevity usually correlates with size; smaller odontocetes such as harbor porpoises are thought to live only for about 15 years. Similarly, the smallest rorqual, the minke whale, lives for about 40 or 50 years, while fin and blue whales may reach 85 to 90 years (Evans, 1987).

Natural Mortality

The deep sea is a poor laboratory setting for studying the full range of factors associated with natural mortality in cetaceans. We recognize predation, disease-causing agents, and, more recently, natural toxins. Cetaceans have a few notable predators: killer whales and sharks. Most of our information on disease is derived from stranded specimens. Virtually every cetacean is parasitized by the time it is weaned. Some of the parasites are capable of causing illness, especially those that migrate to vital organs such as the brain and pancreas (Geraci and St. Aubin, 1987). Many bacteria and some viruses can infect a cetacean, but none in epidemic proportions. Recently, an elusive killer came to light when unusual numbers of humpback whales (Geraci *et al.,* 1989) and bottlenose dolphins (Geraci, 1989) were fatally poisoned by consuming fish contaminated with algal toxins.

Rates of mortality for juveniles and adults are similar in most species. In sperm and sei whales, mortality for all age classes is around 6% per year (Ohsumi, 1979; Allen, 1980), whereas for gray whales (Reilly, 1984) and possibly short-finned pilot whales (Kasuya and Marsh, 1984), the rate for juveniles (11%) may be twice that of adults. In some species of *Stenella*, male mortality is 4 to 7% higher than in females (Kasuya, 1976), perhaps because males fight for mates. The prevalence of scars and bite marks on young males attests to the widespread nature of this competition.

Human-induced mortality is shifting from the purposeful whaling of recent decades to incidental kills related to fishing activities. The role of coastal pollution has yet to be defined. Fishing operations take a heavy toll on pelagic spinner, spotted, and common dolphins of the eastern tropical Pacific. These animals die by the tens of thousands yearly in purse-seines set for yellowfin tuna (Perrin, 1986). Dall's porpoises are killed in the North Pacific by salmon drift-net fisheries (Jones, 1984; Jones *et al.,* 1987). On a smaller but still important scale, several odontocetes are on the verge of extinction due to human activities. The vaquita, or Gulf of California harbor porpoise, has suffered heavy losses in nets set for sea bass (Brownell, 1983; Silber, 1988), and the Chinese river dolphin, or baiji, is almost gone, the victim of heavy fishing activity and habitat destruction (Zhou *et al.,* 1982; and personal observation).

Habitat

The seemingly featureless surface of the sea conceals a range of habitats as different from each other as valleys, mountains, and deserts on land. This is particularly true of the nearshore environment, which is strongly influenced by coastal features, river runoff, and tidal flux. But the pelagic seascape is also varied, with thermoclines, haloclines, and currents directing the nature of biologic events. In the tropics, high surface temperature and salinity result in a stable, stratified water column; in polar regions, ice dominates the physical environment. Such features influence the distribution, migration, and general behavior of cetaceans.

Habitat Use

Whales and dolphins utilize most oceanic habitats. Some habitats, such as coastal shallows or offshore feeding banks, estuaries, bays, nearshore sub-marine canyons, and the ice-edge, might restrict the animals' movements and therefore their ability to avoid contaminated waters. Bowheads, gray whales, belugas, and narwhals are particularly vulnerable to exposure as they pass through narrow leads in heavy ice. Right, humpback, and fin whales in the North Atlantic concentrate much of their activity on the continental shelf to within several kilometers of shore. Right whales winter and apparently calve in coastal waters from Cape Hatteras, North Carolina, to Miami, Florida, with peak abundance in January through March (Winn, 1984; Kraus *et al.,* 1984; Kraus, 1985). From March to April, they live in and around Cape Cod Bay, where they feed and may also calve (W. A. Watkins, personal communication). Thereafter, through November, they concentrate in the Bay of Fundy and the southern Nova Scotian shelf (Mitchell *et al.,* 1986). Humpback and fin whales also occur in shallow waters, nearshore and on banks off Cape Cod, where their concentrations vary with those of sandlance (Hain *et al.,* 1982).

During winter, gray whales of the North Pacific inhabit coastal lagoons and bays of Baja California. During spring and fall they migrate close to shore between Mexico and Alaska, and in summer feed on infaunal benthic invertebrates in coastal shallows of the northern Bering and southern Chukchi Seas (Oliver *et al.,* 1983; Jones and Swartz, 1984; Mate and Harvey, 1984).

Bowhead whales in the western Arctic are similarly constrained. They winter along the ice edge, in open leads, and in polynyas in the mid- to northern Bering Sea, migrate through the Bering Strait and into the Chukchi and Beaufort seas in spring, and spend the summer feeding in the Beaufort Sea (Braham *et al.,* 1984). Summer habitat varies within and between seasons, depending on winds and currents which affect the distribution of copepods and euphausiids. Thus it is not possible to predict their precise location within the summer range (Richardson *et al.,* 1987). Little is known about their distribution patterns along the ice edge in the Bering Sea in winter.

Killer whales, belugas, bottlenose dolphins, and harbor porpoises range close to shore and in restrictive bays and waterways. Their ranges broadly overlap areas of oil production and shipping activities. Some examples are killer whales in Washington State, British Columbia, and the Gulf of Alaska; bottlenose dolphins in the Gulf of Mexico; and beluga whales and harbor porpoises in northern bays and fjords. Other more pelagic toothed whales and dolphins may encounter spilled oil from offshore ship traffic, or when following deep water channels close to shore

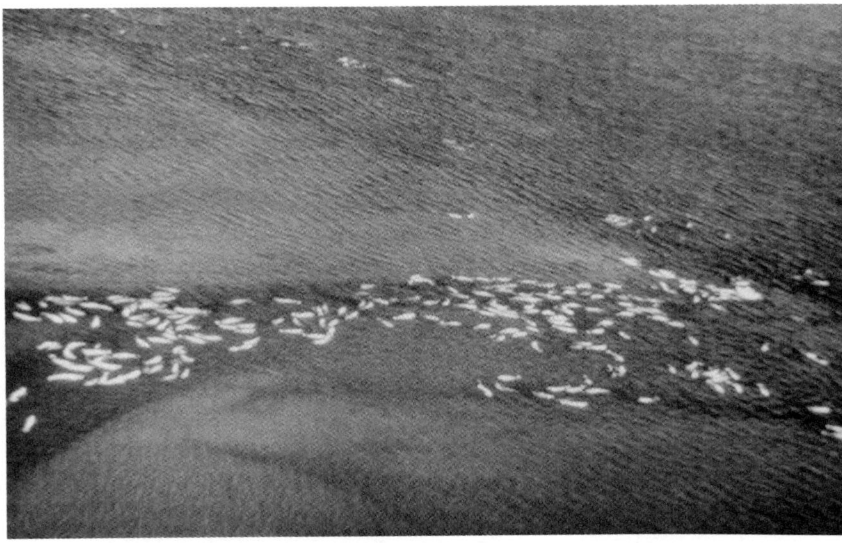

Beluga whales seek the shallows and warmth of estuaries throughout the Arctic during summer. Many hundreds can be observed in Cunningham Inlet on the north shore of Somerset Island, Canada. Photo by T. G. Smith.

in order to feed, as do white-sided dolphins in the Bay of Fundy or Dall's porpoises in the Northeast Pacific.

Home Range and Site Fidelity

Strong attraction to specific areas may override any tendency for a cetacean to avoid the noxious stimuli of oil. Few cetaceans have truly restricted ranges, although in many areas, populations may stay in one general region year round.

Bottlenose dolphins generally have limited home ranges. Dolphins of the Sarasota-Bradenton area of west Florida are year-round residents (Wells *et al.,* 1980; Wells, 1986). While working in Argentina, Melany Würsig and I came to know the movements of a group well enough to predict with some certainty their location off various cliffs and where they were likely to change direction during their traverse through the area (Würsig, 1978). Connor and Smolker (1985) have recently been studying a group of bottlenose dolphins on the west shore of Australia. A part of that population has become habituated to contact with humans, and several generations of dolphins have come to the same small beach to interact with humans for well over 20 years.

Resident killer whales of the Northeast Pacific also show remarkable site fidelity (Bigg, 1982), whereas killer whales of the North Atlantic undergo distinct migrations as they follow herring schools (Jonsgard and Lyshoel, 1970). Individual minke whales have also been seen in the same areas year after year in Puget Sound (Dorsey, 1983). Fin whales may reside in the central Gulf of California year-round (Tershy *et al.*, 1990), and individual right whales have been resighted in the same bays of Argentina over a span of 18 years to date (R. Payne, personal communication). Beluga whales are so strongly attracted to certain estuaries that they return year after year despite vigorous hunting pressure (Caron, 1987).

Most species have the flexibility to search out new areas when old ones become unusable. But whether other habitat is available depends on the geographic extent of the disturbance that is keeping the animals from a particular site. For example, site-tenacious cetaceans such as minke whales of Puget Sound would probably be able to adapt if only one restricted bay or channel were affected, but reproductive success of gray whales would be reduced by a disturbance that affected all avenues to their preferred calving site, the lagoons of Baja California.

Migrations

Many cetacean species migrate, potentially exposing them to spilled oil along the route. Since migration is a necessary event in their life history, oil in their path may modify but is not likely to override the impulse (Evans, 1982b). The classic examples of long-distance migrators are gray, bowhead, right, and humpback whales (Gaskin, 1982). Beyond the general trend to move between feeding and calving grounds, there are important differences in the migration patterns of various species and populations, and even of segments within a population. For example, cued by day lengths of less than 8.5 hours, female humpback whales with newborn calves are last to arrive at and first to leave Antarctic feeding grounds (Dawbin, 1966). Pregnant females spend the longest time on the feeding grounds and do not leave until day length is 6.5 hours or less.

Similarly, migrating gray whales are typically segregated according to age, sex, and reproductive status (Rice and Wolman, 1971). First to reach the calving lagoons are females with near-term fetuses, followed by those which have ovulated and are presumed to have recently weaned a calf. The procession continues with immature females, adult males, and finally immature males. Such spatial and temporal segregation would expose different segments of the population to oil at any given time. Clearly, the greatest concern would be for pregnant females, because of their vital contribution to that year's production, and for the young calves.

Migratory habits of toothed whales are less well defined. Some, such as bottlenose dolphins along the Texas coast (Shane, 1980; Gruber, 1981), Pacific white-sided dolphins off California (Leatherwood and Walker, 1982; Leatherwood *et al.,* 1984), and harbor porpoises in the Bay of Fundy (Gaskin *et al.,* 1985), are only partially migratory. That is, some individuals seem to be resident while others show vague inshore–offshore movements. The longest of these migrations are usually limited to dozens or hundreds, instead of thousands, of kilometers.

Trophic Levels and Feeding Habits

Most mysticetes generally have a restricted diet. Gray whales feed mainly on bottom-dwelling tube worms and ampeliscid amphipods (Nerini, 1984); bowhead and right whales consume clouds of euphausiid crustaceans and calanoid copepods (Lowry and Frost, 1984); whereas smaller rorquals prefer schooling fishes such as menhaden, (*Brevoortia tyrannus*), capelin (*Mallotus villosus*) and sandlance (*Ammodytes* sp.), the larger rorquals prefer euphausiids (Pivorunas, 1979). However, most species are opportunistic and will take advantage of a wide range of prey.

A restricted diet is characteristic of some toothed whales as well, particularly sperm and beaked whales, which feed mainly on squid. Coastal and oceanic dolphins have more catholic diets; their prey selection may vary seasonally between different populations of the same species, and between feeding sessions throughout the day. For example, Hawaiian spinner dolphins, Pacific white-sided dolphins, and pilot whales feed on a variety of mesopelagic fishes and squid. In Argentina, dusky dolphins feed on bottom-dwelling fishes in winter and on schools of southern anchovy (*Engraulis anchoita*) during summer (Würsig and Würsig, 1980). Off New Zealand, the same species concentrates on nonschooling mesopelagic prey throughout the year, since schooling fishes are largely unavailable (Cipriano, 1985; Würsig, 1986).

The diet of toothed whales is not limited to fishes and squid. Bottom-dwelling invertebrates may also be taken by bottlenose dolphins, belugas, and harbor porpoises, among others. Transient killer whales of the Northeast Pacific eat harbor seals and other marine mammals (Heimlich-Boran, 1988); those off Argentina feed on elephant seals, sea lions, and Magellanic penguins (*Spheniscus magellanicus*) (Lopez and Lopez, 1985).

Cetaceans with monotypic diets, or those that depend on seasonally abundant food, might be affected by destruction of local stocks. However, most species of cetaceans are capable of making adjustments in their diet. Therefore, the occurrence and magnitude of nutritional effects will depend on the intensity and geographic coverage of oil and the amount of detrimental impact on alternative prey species. Increased competition for remaining prey may particularly affect species that do not habitually move great distances, such as harbor porpoises, perhaps minke whales, and several species of nearshore odontocetes such as Hector's dolphins or Commerson's dolphins. Yet, typically sedentary bottlenose dolphins off southern California displayed uncharacteristic behavior during the 1982-1983 El Niño. They responded to the incursion of warm water and food scarcity by going 500 km beyond their range and returning when conditions normalized (Wells *et al.*, 1990).

Cetaceans feed at several trophic levels, each with a specific potential to retain and transfer petroleum hydrocarbon residues. Some benthic invertebrates, such as clams and polychaete worms, tend to concentrate these compounds in their tissues, whereas teleosts, and to a lesser extent, crustaceans, metabolize and rapidly excrete them and thus rarely become heavily contaminated even after a major spill (Neff, this volume, Chapter 1). Food web biomagnification of petroleum hydrocarbons does not occur, and animals feeding at the top of a food chain, such as killer whales, are in fact less likely to ingest oil-contaminated food than baleen whales feeding on planktonic or benthic crustaceans. Gray whales (Nerini, 1984), and to some extent bowheads (Lowry and Burns, 1980), face an additional exposure while feeding on bottom sediments where oil may eventually deposit. The potential for ingesting contaminated food is therefore greatest for benthic-feeding whales, less so for planktonic feeders, and least for odontocetes that prey at high trophic levels.

Reproduction

The prospect that oil might disrupt reproductive behavior is remote for offshore species; it is more likely for those inshore. Gray whales give birth and nurture their newborn along the California–Baja California shoreline (Rice and Wolman, 1971). The same pattern exists for humpback whales in Hawaii and the Caribbean, bowheads along the shores of the Bering and Beaufort seas, and right whales in shallow bays of Argentina (Payne, 1980; Payne and Dorsey, 1983) and along the North American coastline (Kraus *et al.*, 1986). The reason for their attraction to warm coastal waters is unclear. Here they are sheltered from oceanic swells and storms, and may be relatively free of large predatory sharks and killer whales. It has also been suggested (Payne and Dorsey, 1983) that mothers and calves retreat to shallow waters to evade the boisterous mating attempts of males. Because births of baleen whales are usually sharply seasonal, disruption of preferred habitat could interfere with normal reproduction. Shallow water is also an important breeding habitat for some odontocetes, such as spinner dolphins, bottlenose dolphins, harbor porpoises, beluga whales, and narwhals. The latter two concentrate breeding and calving activities within a very short season and therefore are more vulnerable to disruption at these times, but less likely to be disturbed during most of the year.

Social Organization and Behavior

Group Structure and Social Behavior

Group structure is fundamentally different in baleen whales and toothed whales. The former tend to migrate, feed, and breed as loose aggregations, with individuals often separated by hundreds of meters. Mother–calf pairs excepted, their social bonds last only hours or days. Members of such loose aggregations probably communicate and may react collectively to a threat. Southern right whales huddle close together, all flailing their tails at the approach of killer whales (R. Payne, personal communication), and bowhead whales are reported to behave similarly when startled by seismic blasts (Reeves *et al.,* 1984). During the early days of hunting (Scammon, 1874), whalers observed the same reaction to the presence of their own vessels. During such times, social cohesion may be greater than is generally supposed for baleen whales.

Recent data on balaenopterid whales suggest that general social organization may be strongly influenced by feeding mode. Thus, blue and fin whales, which are plankton feeders, tend to range over large areas while searching for sporadically abundant prey. Smaller balaenopterids, such as Bryde's and minke whales, are more resident while feeding on schools of fishes that may be locally abundant. Whereas the larger species of whales tend to travel in loose social units of several animals, Bryde's and minke whales are more solitary (in most nearshore localities), possibly in order to avoid intraspecific competition (Tershy and Breese, 1987).

In sharp contrast to the loose aggregations typical of baleen whales, most toothed whales are gregarious and usually live in relatively stable groups, in which bonds of kinship are probably secure. This is certainly so for killer whales of Puget Sound and British Columbia (Bigg, 1982; Heimlich-Boran, 1986; Bigg *et al.,* 1987), bottlenose dolphins (Würsig, 1978; Shane *et al.,* 1986; Wells, 1986), and many other coastal and pelagic dolphin species (Morozov, 1970; Norris and Dohl, 1980; Wells *et al.,* 1980; Würsig and Würsig, 1980; Cipriano, 1985). Schooling behavior could result in large numbers of animals coming in contact with oil during a spill. Because odontocetes in groups are constantly communicating, enhanced sensory integration may also allow them to more efficiently detect oil, and therefore avoid it as a group.

Cetaceans communicate by sound, which has the advantage (in water) of allowing them to interact over a far greater range than would be possible using vision alone. Even clear "blue-water" allows for visual contact within only 100 m or so, whereas some auditory signals are rapidly propagated, with little attenuation, for dozens of kilometers. Hearing is especially useful in murky water, at night, and at depth.

Baleen whales produce mainly low and medium frequency sounds, from about 10 Hz to about 3000 Hz (Schevill *et al.,* 1964). Humpback whales generate complex, variable notes (Silber, 1986), strung together to form songs [in the proper definition of songs as repeating sequences of notes (Payne and McVay, 1971)] that are quite melodic. Gray whales grunt and click, and right and bowhead whales moan (Clark, 1983; Clark and Johnson, 1984). These sounds may be used to maintain group cohesion and spacing, facilitate social and reproductive activities, and signal alarm. None, however, is within the frequency range for echolocation. Possibly the clicks made by gray whales provide them with information about the shoreline and other gross features of the environment.

In odontocetes, communication-type signals tend to consist of whistles and rapid successions of clicks which may sound to us like creaks or rasps. Some species also use ultrasonic clicks, up to 200 kHz (over 10 times the upper level of hearing for humans), to echolocate. This highly discriminating sense is used to find food, investigate objects, "map" the environment, and detect other dolphins and potential danger. Dolphins do not echolocate at all times, however, and during rest and low-level socializing they can be remarkably silent. Their large predatory cousin, the killer whale, is also silent for long periods, and one gets the impression that killer whales and dusky dolphins of Argentina, for example, play an elaborate cat and mouse game as they manoeuvre within several kilometers of each other. Sight is probably of great importance at such silent times.

Group structure and social behavior vary diurnally and seasonally, at least for dolphins that habitually come within sight of shore. For example, dusky dolphins off Argentina spend winter in groups of about 10 animals within 1 km of shore as they feed on bottom-dwelling prey. In spring, summer, and fall, however, they hunt in small groups for inshore schooling fish in the morning, and aggregate temporarily into large, offshore feeding and socializing groups of up to 300 animals around mid-day (Würsig and Würsig, 1980). Groups of 10–100 Hawaiian spinner dolphins rest in protective bays in the daytime, and at night aggregate to feed on mesopelagic fishes and squid 2–10 km from shore (Norris *et al.,* 1990). Similarly, dusky dolphins off New Zealand feed on mesopelagic fishes and squid in deep water at night and rest close to shore in the daytime (Würsig *et al.,* 1990). However, this pattern is more pronounced off New Zealand in spring and summer than in the seasonally uniform Hawaiian waters. It is clear that risk of exposure to oil can change considerably as a function of diurnal and seasonal changes in behavior.

The composition of social groups is not necessarily uniform by age and sex. As a result, segments of the population may be differentially exposed to oil. Gray and right whale mothers and calves tend to segregate themselves in nursing areas, and gray whales do so during migration as well (Rice and Wolman, 1971). Sperm whale males travel either alone or in bachelor herds, often far from females and young (Best, 1979). Pilot whales may also form age and sex-segregated groups, as may bottlenose dolphins (Wells *et al.*, 1980) and dusky dolphins (Würsig and Würsig, 1980) that live in temporary societies or subgroups. In the latter species, segregation may occur at feeding times.

In spite of obvious advantages, social cohesion has its drawbacks. Some odontocetes, notably pilot whales, false killer whales, white-sided dolphins, and sperm whales, remain together until they die in mass strandings (Geraci and St. Aubin, 1979). A herd approaching an oil spill would be expected to react as a unit, with great probability that the last one to respond to the oil would do so as the first.

Reproductive Behavior

Disruption of herd structure (as by oil) may affect elements of the social order that ensure reproductive success. In many odontocete species, attendant females have important nurturant roles to female-calf pairs (Caldwell and Caldwell, 1972). These so-called "auntie" dolphins assist during parturition, at times apparently helping to lift the calf to the surface and supporting it there (McBride and Kritzler, 1951). Females cooperate to supervise calves, allowing each other in turn to forage away from the herd (Tavolga and Essapian, 1957). When threatened by aggressive males, older females protect calves by herding them to the center of the school.

Aggression features prominently in the establishment and maintenance of social hierarchies, and in turn governs the reproductive activities of the males. Norris and Prescott (1961) commented on the high frequency of scars, presumably inflicted by older bulls, in young male Pacific pilot whales. Among bottlenose dolphins, dominant males direct their aggression toward females that have withdrawn from the herd to calve (McBride and Kritzler, 1951). The integrity of the social order likely contributes to the collective safety of the group and, as such, confers survival value. Disruption of the hierarchy, by oil or any other significant environmental disturbance, might trigger a cascade of events affecting mothers, calves, or vital social bonds.

Sexual behavior in mysticetes is sharply seasonal, with mating and calving occurring in the same season and area (generally winter or spring).

It is also seasonal in those toothed whales that live in, or migrate to, higher latitudes, such as beluga whales, narwhals, and sperm whales. However, many social odontocetes are sexually active throughout the year, though coupling for procreation for one season only. That is, sexual behavior appears to be an important aspect of social behavior, perhaps related to bonding and the establishment and maintenance of social hierarchies. Dolphins in large schools are composed of stable subgroups which may represent family units (at least with mothers and their female relatives) to some degree, but with social and sexual affiliations of members both within and among subgroups (Wells, 1986, provides a good summary). Dolphin societies thus show a degree of social "openness" between subgroups, not unlike some neotropical spider monkeys (Würsig and Würsig, 1979, 1980; Norris *et al.*, 1990).

Feeding Behavior

Cetaceans use a variety of feeding strategies, as might be expected from the diversity of their prey. Because of differences in food dispersion patterns, mysticete fish-eaters tend to be more solitary than those feeding on krill (Tershy and Breese, 1987). When feeding on schooling fishes, dolphins and toothed whales often coordinate their activities to herd and contain prey. Apparently organized attacks by killer whale pods on seals (Smith *et al.*, 1981) and a blue whale (Tarpy, 1979) have been described. Baleen whales generally feed alone, although apparent cooperation during feeding has recently been described in humpback (Baker and Herman, 1985; D'Vincent *et al.*, 1985), fin (Tershy *et al.*, 1990), right (S. D. Kraus, personal communication), and bowhead (Würsig *et al.*, 1985) whales.

Especially dramatic is the coordinated hunting by humpback whales, which will lunge together at the surface in groups of up to 22 whales (T. Jefferson, personal communication and photographs from observations in southern Alaska). Humpback whales also create "bubble nets" which encircle prey long enough for the whales to engulf them. Several neighboring whales may bubble-net at the same time, although whether by design or coincidence is not clear (Jurasz and Jurasz, 1979; Hain *et al.*, 1982). All of these feeding strategies tend to keep whales at the surface where there is greater likelihood of repeated exposure to oil.

How a cetacean consumes its food affects the likelihood of its ingesting oil. This is particularly true of mysticetes, which have an added potential for fouling the baleen, the food-gathering apparatus. To appreciate how fouling might occur, it is important to understand the design and function of this unique structure. All baleen whales feed by taking small invertebrates or fish into the mouth and expelling water through baleen plates which hang from the upper jaw. These plates are remarkable adaptations and are the single most distinguishing feature of the mysticetes, or "moustache whales." The plates are composed of keratin, similar to mammalian claws and hair. Leatherwood and Reeves (1983) give an apt description:

> *Each of the plates forms roughly a right triangle, with the long vertical edge oriented to face outside the mouth, the base embedded in the gum, and the hypotenuse inside the mouth. The outer edge is smooth; the inner edge is frayed, and consists of a series of bristles or "hairs", which differ in appearance and number from one species to another, and can be used to distinguish them. The plates are arranged somewhat like the teeth in a comb, one row along each side of the mouth. Inside the mouth the bristles intertwine to form a sieve or mat. (p. 2–3)*

Each plate is separated from the next by 1–2 cm, and one side of the upper jaw may house (in the bowhead) as many as 360 plates. The bowhead also has the longest (up to 4 m), most flexible, and most finely fringed baleen. As the whale swims, water enters the mouth and is passed through the baleen curtain on either side, leaving the food in the mouth. The closely-spaced plates and fine mesh of hairs allow the Balaenidae to feed on copepods as small as several millimeters in length.

The rorquals are the most active feeders, often rapidly lunging through clouds of prey, taking in huge gulps of water and prey while distending their throats, then partially closing the mouth and squirting water through short, coarse baleen. Rorquals have a series of skin folds or pleats on the undersurface of the lower jaw enabling them to take in large volumes of water. These extend from the tip of the jaw to as far back as the umbilicus, and they distend, much as an accordian can open up, to provide a cavernous throat which may take in tons of water and prey in one gulp. The distention is made possible not only by the pleats, but by a highly elastic tongue which can spread out and line the throat cavity (Lambertsen, 1983). Thus water is pumped out by the closing of the jaws, the motion of the tongue, and when a whale gulps food near the surface, by bringing part of the mouth above the water line and emptying the throat. Meanwhile, the mouth is closed tightly enough to force water through the short baleen plates and thus keep the prey contained. In this active lunging and gulping fashion, rorquals take in both invertebrates and fish. Here again, external morphology is adapted to feeding mode, for the rorquals appear well built for bursts of speed necessary to gulp clouds of prey (especially the swifter fishes) with their relatively small mouths.

The gray whale has an unusual feeding strategy for obtaining invertebrates from bottom sediments. While moving on its side, it sucks up organisms in pulses, leaving a trail of oblong mouth-sized depressions in the sediment where prey have been removed (Nerini, 1984).

Pivorunas (1979) suggested that each major group of mysticetes has a characteristic feeding strategy: right whales skim the surface, rorquals gulp in the water column, and gray whales scour the bottom. However, more detailed studies are showing that there is considerable overlap in the way they gather their food. Bowheads also filter prey just above bottom sediments (Würsig *et al.*, 1985), rorquals break the surface at the end of a feeding lunge (Hain *et al.*, 1982), and gray whales eat pelagic fish and mysids in the water column (Rice and Wolman, 1971; Murison *et al.*, 1984). Thus, we can only hypothesize that whales feeding in the water column are less apt to be exposed to oil than are those feeding at the surface or on the bottom.

The same would be true of odontocetes, having equally diverse feeding strategies. Most species probably feed within the water column, though some, such as belugas, exploit benthic organisms (Vladykov, 1946). Dolphins that toss fish through an oiled surface or work as a group to herd them there, risk contaminating their prey.

Other Behaviors

Whales and dolphins are curious and will play with objects at the surface. Hawaiian spinner dolphins approach floating plastic strips washed to sea from pineapple fields (Norris *et al.*, 1990); dusky dolphins may play with

A trail of mud marks the path of a feeding gray whale as it filters organisms from sediments gathered on the bottom. Photo by B. Würsig.

loose strands of kelp for long periods of time (Würsig and Würsig, 1979); and bowhead whales play with floating logs and other debris (Würsig *et al.*, 1985, 1989). A bowhead calf was observed orienting to, and playing within a surface area fouled by the chemical marker fluorescein. Another calf oriented along a long line of surface debris caused by a current boundary and repeatedly opened its mouth while swimming through the debris (Würsig *et al.*, 1985). The latter incidents occurred while the mother was below the surface, probably feeding, and the calf was left alone to explore on its own. These observations illustrate what is known for young mammals in general: they are especially curious and are prone to mischief when a parent is not present. Bowheads probably are not the only mysticetes to leave their calves in this manner, though others have not been adequately studied to prove this point. Deep-diving odontocete species, such as sperm whales, pilot whales, and some pelagic dolphins, may also leave their calves behind, as may Dall's (Jefferson, 1987) and harbor (Amundin and Amundin, 1974; Gaskin and Watson, 1985) porpoises. In such situations, curious calves could be drawn to oil slicks or emulsions on the sea surface.

Summarizing the Risk

Many aspects of behavior, diet, and habitat use may lead cetaceans into contact with spilled oil. Given the host of interacting variables, it is difficult to state precisely which species or individuals might be most vulnerable. However, several assumptions can be made. These are summarized in Table 5–2.

Encounters with oil are likely to be prolonged in species that frequent restricted areas such as bays and estuaries. Examples are breeding and feeding humpback, gray, right, bowhead, and beluga whales, narwhals, bottlenose dolphins, harbor porpoises, and river dolphins. Cetaceans that range widely may contact some oil as they move quickly through a fouled area, but are less likely to suffer long-term exposure.

After a spill, oil is distributed primarily at the surface and on the seabed (Neff, this volume, Chapter 1). Consequently, cetaceans that feed in these areas are more likely to contact oil than those that feed in the water column. These include skim-feeding right and bowhead whales, surface-lunging rorquals, and bottom-feeding gray whales—in other words, all mysticetes, except possibly the minke whale. Harbor porpoises and some dolphins may contact oil when they feed on flatfish and other bottom-dwelling prey. Dolphins that habitually force schools of prey to the surface may also be at risk.

Table 5-2
Cetacean Vulnerability to Oil as a Function of Life History, with Subjective Rankings from 1 (Low Risk) to 5 (High Risk)

	Criteria				
Family/species	Range[a]	Habitat[b]	Prey diversity[c]	Behavioral flexibility[d]	Population[e]
Balaenopteridae					
Blue	3	3	4	3	5
Fin	3	3	4	3	4
Sei	2	2	3	3	3
Bryde's	2	2	3	3	3
Humpback	3	5	3	2	4
Minke	4	4	3	3	2
Balaenidae					
N. & S. Right	4	5	4	3	5
Pygmy Right	2	2	3	3	4
Bowhead	4	5	4	3	5
Eschrichtiidae					
Gray	3	5	5	4	2
Physeteridae	1	1	3	2	3
Monodontidae					
White Whale	3	5	3	3	2
Ziphiidae	1	1	3	3	3
Delphinidae					
Pelagic	1	1	2	4	2
Coastal	5	5	2	2	4
Platanistidae	4	5	3	4	5
Phocoenidae (all but 2 below)	5	5	3	4	5
Dall's porpoise	2	2	3	4	3
Finless porpoise	3	4	3	4	3

[a] Large range (1) to small range (5).
[b] Oceanic (1) to coastal, ice dwelling, riverine (5).
[c] Diverse prey (1) to limited diet (5).
[d] Adaptable (1) to sensitive to stress (5).
[e] Large population (1) to endangered (5).

As a group, baleen whales appear to be the most vulnerable in view of their generally low numbers, their peculiar feeding strategies, and their dependence on selected, localized habitats for feeding and reproduction. Among the odontocetes in North American waters, restrictive habitat increases the risk of exposure for belugas, narwhals, harbor porpoises, and bottlenose dolphins. Most other odontocetes are too mobile and wide-ranging for oil to present much of a threat.

Acknowledgments

Sheila Baldridge (Moss Landing Marine Laboratories) and Thomas Jefferson (Texas A&M University at Galveston) assisted in the preparation of this manuscript. This represents Contribution No. 6 of the Marine Mammal Research Program, Texas A & M University at Galveston.

References

Allen, K. R. (1980). "Conservation and Management of Whales." Univ. of Washington, Press, Seattle.

Amundin, M., and Amundin, B. (1974). On the behaviour and study of the harbour porpoise, *Phocoena phocoena,* in the wild. *Invest. Cetacea* **5,** 317–328.

Baker, S. C., and Herman, L. M. (1985). Whales that go to extremes. *Nat. Hist.* **10,** 52–60.

Barnes, L. G. (1984). Whales, dolphins and porpoises: Origin and evolution of the Cetacea. *In* "Mammals: Notes for a Short Course" (T. W. Broadhead, ed.), Stud. Geol., No. 8. University of Tennessee, Knoxville.

Berzin, A. A. (1984). Soviet studies of the distribution and numbers of the gray whale in the Bering and Chukchi Seas, from 1968 to 1982. *In* "The Gray Whale, *Eschrichtius robustus*" (M. L. Jones, S. L. Swartz, and S. Leatherwood, eds.), pp. 409–419. Academic Press, Orlando, Florida.

Best, P. B. (1979). Social organization of sperm whales, *Physeter macrocephalus. In* "Behavior of Marine Animals" (H. E. Winn and B. L. Olla, eds.), Vol. 3, pp. 227–289. Plenum, New York.

Best, P. B., Canham, P. A. S., and MacLeod, N. (1984). Patterns of reproduction in sperm whales, *Physeter macrocephalus. Rep. Int. Whal. Comm., Spec. Issue* **6,** 51–80.

Bigg, M. A. (1982). An assessment of killer whale (*Orcinus orca*) stocks off Vancouver Island, B.C. *Rep. Int. Whal. Comm.* **32,** 655–666.

Bigg, M. A., Ellis, G. M., Ford, J. K. B., and Balcomb, K. C. (1987). "Killer Whales." Phantom Press, Nanaimo, B.C.

Braham, H. W. (1984). The status of endangered whales: An overview. *Mar. Fish. Rev.* **46**(4), 2–6.

Braham, H. W., Krogman, B. D., and Carroll, G. M. (1984). Bowhead and white whale migration, distribution and abundance in the Bering, Chukchi and Beaufort Seas, 1975-1978. *NOAA Tech. Rep.* **NMFS SSRF-778,** 1–39.

Brown, S. G., and Lockyer, C. H. (1984). Whales. *In* "Antarctic Ecology" (R. M. Laws, ed.), Vol. 2, pp. 717–781. Academic Press, Orlando, Florida.

Brownell, R. L., Jr. (1983). *Phocoena sinus. Mamm. Species* **198,** 1–3.

Brownell, R. L., Jr., Ralls, K., and Perrin, W. F. (1989). The plight of the 'forgotten' whales. *Oceanus* **32**(1), 5–13.

Buckland, S. T., and Anganuzzi, A. A. (1988). Estimated trends in abundance of dolphins associated with tuna in the eastern tropical Pacific. *Rep. Int. Whal. Comm.* **38,** 411–437.

Caldwell, D. K., and Golley, F. B. (1965). Marine mammals from the coast of Georgia to Cape Hatteras. *J. Elisha Mitchell Sci. Soc.* **81,** 24–32.

Caldwell, M. C., and Caldwell, D. K. (1972). Behavior of marine mammals. *In* "Mammals of the Sea: Biology and Medicine" (S. H. Ridgway, ed.), pp. 419–465. Thomas, Springfield, Illinois.

Caron, L. M. J. (1987). Status, site fidelity, and behavior of a hunted herd of white whales, *Delphinapterus leucas* in the Nastapoka Estuary, eastern Hudson Bay. M.Sc. Thesis, Dept. of Renewable Resources, McGill University, Montreal, Canada.

Cipriano, F. (1985). Dusky dolphin research at Kaikoura, New Zealand: A progress report. *Mauri Ora* **12,** 151–158.

Clark, C. W. (1983). Acoustic communication and behavior of the southern right whale (*Eubalaena australis*). *In* "Communication and Behavior of Whales" (R. Payne, ed.), pp. 163–198. Westview Press, Boulder, Colorado.

Clark, C. W., and Johnson, J. H. (1984). The sounds of the bowhead whale, *Balaena mysticetus*, during the spring migrations of 1979 and 1980. *Can. J. Zool.* **62,** 1436–1441.

Connor, R. C., and Smolker, R. S. (1985). Habituated dolphins (*Tursiops* sp.) in western Australia. *J. Mammal.* **66,** 398–400.

Dawbin, W. H. (1966). The seasonal migratory cycle of humpback whales. *In* "Whales, Dolphins, and Porpoises" (K. S. Norris, ed.), pp. 145–170. Univ. of California Press, Berkeley.

Dohl, T. P., Guess, R. C., Duman, M. L., and Helm, R. C. (1983). Cetaceans of central and northern California, 1980–83. *In* "Book 2: Status, Abundance and Distribution," Vol. 3, Final Rep. No. 84-0045. U.S. Minerals Manage. Serv., University of California, Santa Cruz.

Dohl, T. P., Bonnell, M. L., and Glenn Ford, R. (1986). Distribution and abundance of common dolphin, *Delphinus delphis*, in the southern California Bight: A quantitative assessment based upon aerial transect data. *Fish. Bull.* **84,** 333–343.

Dorsey, E. M. (1983). Exclusive adjoining ranges in individually identified minke whales (*Balaenoptera acutorostrata*) in Washington state. *Can. J. Zool.* **61,** 174–181.

D'Vincent, C. G., Nilson, R. M., and Hanna, R. E. (1985). Vocalization and coordinated feeding behavior of the humpback whale in southeastern Alaska. *Sci. Rep. Whales Res. Inst.* **36,** 41–47.

Evans, P. G. H. (1987). "The Natural History of Whales and Dolphins." Facts on File, New York.

Evans, W. E. (1982a). Distribution and differentiation of stocks of *Delphinus delphis* Linnaeus in the northeastern Pacific. *In* "Mammals in the Seas," Vol. 4., pp. 45–66. Fish. Ser. No. 5., Food Agric. Organ. United Nations, Rome.

Evans, W. E. (1982b). A study to determine if gray whales detect oil. *In* "Study of the Effects of Oil on Cetaceans" (J. R. Geraci and D. J. St. Aubin, eds.), Final Rep., Contract No. AA 551-CT9-29, pp. 47–61. U.S. Dept. of the Interior, Bur. Land Manage., Washington, D. C.

Gaskin, D. E. (1982). "The Ecology of Whales and Dolphins." Heinemann Press, London.

Gaskin, D. E., and Watson, A. P. (1985). The harbor porpoise, *Phocoena phocoena,* in Fish Harbour, New Brunswick, Canada: Occupancy, distribution, and movements. *Fish Bull.* **83,** 427–442.

Gaskin, D. E., Arnold, P. W., and Blair, B. A. (1974). *Phocoena phocoena*. *Mamm. Species* **42,** 1–8.

Gaskin, D. E., Smith, G. J. D., Watson, A. P., Yasui, W. Y., and Yurick, D. B. (1984). Reproduction in the porpoises (Phocoenidae): Implications for management. *Rep. Int. Whal. Comm., Spec. Issue* **6,** 135–148.

Gaskin, D. E., Read, A. J., Watts, P. F., and Smith, G. J. D. (1985). Population dispersal, size, and interactions of harbour porpoises in the Bay of Fundy and Gulf of Maine, *Can. Tech. Rep. Fish. Aquat. Sci.* **1291,** 1–28.

Geraci, J. R. (1989). "Clinical Investigation of the 1987-88 Mass Mortality of Bottlenose Dolphins Along the U.S. Central and South Atlantic Coast," Final Rep. U.S. Nat. Mar. Fish. Serv., Off. Nav. Res., Mar. Mammal Comm., Washington, D.C.

Geraci, J. R., and St. Aubin, D. J., eds. (1979). "Biology of Marine Mammals: Insights through Strandings," PB293-890. Natl. Tech. Inf. Serv., Springfield, Virginia.

Geraci, J. R., and St. Aubin, D. J. (1987). Effects of parasites on marine mammals. *Int. J. Parasitol.* **17,** 407–414.

Geraci, J. R., Anderson, D. M., Timperi, R. J., St. Aubin, D. J., Early, G. A., Prescott, J. H., and Mayo, C. A. (1989). Humpback whales (*Megaptera novaeangliae*) fatally poisoned by dinoflagellate toxin. *Can. J. Fish. Aquat. Sci.* **46,** 1895–1898.

Gruber, J. A. (1981). Ecology of the Atlantic bottlenosed dolphin (*Tursiops truncatus*) in the Pass Cavallo area of Matagorda Bay, Texas. M.Sc. Thesis, Texas A&M Univ., College Station.

Hain, J. H. W., Carter, G. R., Kraus, S. K., Mayo, C. A., and Winn, H. E. (1982). Feeding behavior of the humpback whale, *Megaptera novaeangliae*, in the western North Atlantic. *Fish. Bull.* **80**(2), 259–268.

Hall, I. D. (1979). "A Survey of Cetaceans of Prince William Sound and Adjacent Vicinity—Their Numbers and Seasonal Movements," Rep. Contract No. RU-481. U.S. Minerals Manage. Serv., Reston, Virginia.

Heimlich-Boran, S. L. (1986). Cohesive relationships among Puget Sound killer whales. *In* "Behavioral Biology of Killer Whales" (B. C. Kirkevold and J. S. Lockard, eds.), pp. 251–284. Liss, New York.

Heimlich-Boran, J. R. (1988). Behavioral ecology of killer whales in the Pacific northwest. *Can. J. Zool.* **66,** 565–578.

Heyning, J. E., and Dahlheim, M. E. (1988). *Orcinus orca. Mamm. Species* **304,** 1–9.

International Whaling Commission (IWC) (1986). Report of the Scientific Committee, Annex I (Small Cetaceans). *Rep. Int. Whal. Comm.* **36,** 112–117.

International Whaling Commission (IWC) (1987). Report of the Sub-Committee on Small Cetaceans, Annex H. *Rep. Int. Whal Comm.* **37,** 121–126.

Jefferson, T. A. (1987). A study of the behavior of Dall's porpoise *Phocoenoides dall;* in the Johnstone Strait, British Columbia. *Can. J. Zool.* **65,** 736–744.

Jefferson, T. A. (1988). *Phocoenoides dalli. Mammal. Species* **319,** 1–7.

Jones, L. L. (1984). Incidental take of the Dall's porpoise and the harbor porpoise by Japanese salmon driftnet fisheries in the western North Pacific. *Rep. Int. Whal. Comm.* **34,** 531–538.

Jones, L. L., Bouchet, G. C., and Turlock, B. J. (1987). "Comprehensive report on the incidental take, biology, and status of Dall's porpoise." Unpublished report NMML-NMFS, 7600 Sand Point Way N. E., Seattle, Washington 98115.

Jones, M. L., and Swartz, S. L. (1984). Demography and phenology of gray whales and evaluation of whale–watching activities in Laguna San Ignacio, Baja California Sur, Mexico. *In* "The Gray Whale, *Eschrichtius robustus*" (M. L. Jones, S. L. Swartz, and S. Leatherwood, eds.), pp. 309–374. Academic Press, Orlando, Florida.

Jonsgard, A., and Lyshoel, P. B. (1970). A contribution to the biology of the killer whale, *Orcinus orca* (L.). *Norw. J. Zool.* **18,** 41–48.

Jurasz, C. M., and Jurasz, V. P. (1979). Feeding modes of the humpback whale, *Megaptera novaeangliae*, in southeast Alaska. *Sci. Rep. Whales Res. Inst.* **31,** 69–83.

Kasuya, T. (1976). Reconsideration of life history parameters of the spotted and striped dolphins based on cemental layers. *Sci. Rep. Whales Res. Inst.* **28,** 73–106.

Kasuya, T., and Marsh, H. (1984). Life history and reproductive biology of the short-finned pilot whale *Globicephala macrorhynchus* off the Pacific coast of Japan. *Rep. Int. Whaling Comm., Spec. Issue* **6,** 259–310.

Kraus, S. D. (1985). "A Review of the Status of Right Whales (*Eubalaena glacialis*) in the Western North Atlantic, with a Summary of Research and Management Needs." Final Rep. No. MMC-84/08. U.S. Marine Mammal Comm., Washington, D. C.

Kraus, S. D., Prescott, J. H., and Stone, G. S. (1984). "Right Whales (*Eubalaena glacialis*) in the Western North Atlantic: A Summary of Research Conducted from 1980 to 1984," Rep. No. MMC-442. U.S. Mar. Mammal Comm., Washington, D. C.

Kraus, S. D., Prescott, J. H., and Knowlton, A. R. (1986). "Wintering Right Whales (*Eubalaena glacialis*) Along the Southeastern Coast of the United States, 1984-1986," Rep. U.S. Minerals Manage. Serv., Reston, Virginia.

Lambertsen, R. H. (1983). Internal mechanism of rorqual feeding. *J. Mammal.* **64,** 76–88.

Leatherwood, S., and Reeves, R. R. (1983). "The Sierra Club Handbook of Whales and Dolphins." Sierra Club Books, San Francisco, California.

Leatherwood, S., and Walker, W. A. (1979). The northern right whale dolphin, *Lissodelphis borealis* Peale, in the eastern North Pacific. *In* "Behavior of Marine Mammals" (H. E. Winn and B.L. Olla, eds.), Vol. 3, pp. 85–141. Plenum, New York.

Leatherwood, S., and Walker, W. A. (1982). "Population Biology and Ecology of the Pacific White-Sided Dolphin, *Lagenorhynchus obliquidens*, in the Northeastern Pacific. Part I. Distribution, Seasonal Movements and Abundance, with a Bibliography and Summary of Specimen Material," Admin. Rep. No. LJ-82-18C. Natl. Mar. Fish. Serv., Southwest Fish. Cent., La Jolla, California.

Leatherwood, S., Caldwell, D. K., and Winn, H. E. (1976). Whales, dolphins, and porpoises of the western North Atlantic. *NOAA Tech. Rep., NMFS Circ.* **396.**

Leatherwood, S., Perrin, W. F., Kirby, V. L., Hubbs. C. L., and Dahlheim, M. (1980). Distribution and movements of Risso's dolphin, *Grampus griseus* in the eastern North Pacific. *Fish Bull.* **77,** 951–963.

Leatherwood, S., Reeves, R. R., Perrin, W. F., and Evans, W. E. (1982). Whales, dolphins, and porpoises of the eastern North Pacific and adjacent Arctic waters. *NOAA Tech. Rep., NMFS Circ.* **444.**

Leatherwood, S., Reeves, R. R., Bowles, A. E., Stewart, B. S., and Goodrich, K. R. (1984). Distribution, seasonal movements, and abundance of Pacific white-sided dolphins in the eastern North Pacific. *Sci. Rep. Whales Res. Inst.* **35,** 129–157.

Lockyer, C. (1981). Growth and energy budgets of large baleen whales from the southern hemisphere. *In* "Mammals in the Seas," Vol. 3, Fish. Ser. No. 5, pp. 379–487. Food Agric. Organ. United Nations, Rome.

Lockyer, C. (1984). Review of baleen whale (*Mysticeti*) reproduction and implications for management. *Rep. Int. Whal. Comm., Spec. Issue* **6,** 27–50.

Lopez, J. D., and Lopez, D. (1985). Killer whales (*Orcinus orca*) of Patagonia, and their behavior of intentional stranding while hunting nearshore. *J. Mammal.* **66,** 181–183.

Lowry, L. F., and Burns, J. J. (1980). Food utilized by bowhead whales near Barter Island, Alaska, autumn 1979. *Mar. Fish. Rev.* **42,** 88–91.

Lowry, L. F., and Frost, K. J. (1984). Food and feeding of bowhead whales in western and northern Alaska. *Sci. Rep. Whales Res. Inst.* **35,** 1–16.

Masaki, Y. (1977). The separation of the stock units of sei whales in the North Pacific. *Rep. Int. Whal. Comm., Spec. Issue* **1,** 71–79.

Mate, B. R. (1989). Watching habits and habitats from earth satellites. *Oceanus* **32**(1), 14–18.

Mate, B. R., and Harvey, J. T. (1984). Ocean movements of radio-tagged gray whales. *In* "The Gray Whale, *Eschrichtius robustus*" (M. L. Jones, S. L. Swartz, and J. S. Leatherwood, eds.), pp. 577–590. Academic Press, Orlando, Florida.

Matkin, C. O., and Leatherwood, S. (1986). General biology of the killer whale, *Orcinus orca*: A synopsis of knowledge. *In* "Behavioral Biology of Killer Whales" (B. C. Kirkevold and J.S. Lockard, eds.), pp. 35–68. Liss, New York.

McBride, A. F., and Kritzler, H. (1951). Observations on pregnancy, parturition and postnatal behavior in the bottlenose dolphin. *J. Mammal.* **32,** 251–266.

Meredith, G. N., and Campbell, R. R. (1988). Status of the fin whale, *Balaenoptera physalus*, in Canada. *Can. Field Nat.* **102,** 351–368.

Mitchell, E. (1975). "Porpoise, Dolphin and Small Whale Fisheries of the World: Status and Problems," IUNC Monogra. No. 3. International Union for the Conservation of Nature, Morges, Switzerland.

Mitchell, E., and Reeves, R. R. (1981). Catch history and cumulative catch estimates of initial population size of cetaceans in the eastern Canadian Arctic. *Rep. Int. Whal. Comm.* **31,** 645–682.

Mitchell, E., Kozicki, V. M., and Reeves, R. R. (1986). Sightings of right whales, *Eubalaena glacialis*, on the Scotia shelf, 1966-1972. *Rep. Int. Whal. Comm., Spec. Issue* **10,** 83–107.

Morejohn, G. V. (1979). The natural history of Dall's porpoise in the north Pacific Ocean. *In* "Behavior of Marine Animals" (H. E. Winn and B. L. Olla, eds.), Vol. 3, pp. 45–83. Plenum, New York.

Morozov, D. A. (1970). Dolphins hunting. *Rybn. Khoz.* (Moscow) **46,** 16–17. (transl. by Lillian Vlymen, Natl. Mar. Fish. Serv., La Jolla, California).

Murison, L. D., Murie, D. J., Morin, K. R., and da Silva Curiel, J. (1984). Foraging of the gray whale along the west coast of Vancouver Island, British Columbia. *In* "The Gray Whale, *Eschrichtius robustus.*" (M. L. Jones, S. L., Swartz, and S. Leatherwood, eds.), pp. 451–463. Academic Press, New York.

National Marine Fisheries Service (NMFS) (1988). "Annual Report 1987-88, Marine Mammal Protection Act of 1972," Appendix Table 9. NOAA, Dept. of Commerce, Washington, D. C.

National Marine Fisheries Service (NMFS) (1987). U.S. progress report on cetacean research, June 1985 to May 1986. *Rep. Int. Whal. Comm.* **37,** 183–190.

Nerini, M. K. (1984). A review of gray whale feeding ecology. *In* "The Gray Whale, *Eschrichtius robustus*" (M. L. Jones, S. L. Swartz, and S. Leatherwood, eds.), pp. 423–450. Academic Press, Orlando, Florida.

Nerini, M. K., Braham, H., Marquette, W., and Rugh, D. (1984). Life history of the bowhead whale. *J. Zool.* **204,** 443–468.

Nishiwaki, M. (1975). Ecological aspects of smaller cetaceans, with emphasis on the striped dolphin (*Stenella coeruleoalba*). *J. Fish. Res. Board Can.* **32,** 1069–1072.

Norris, K. S., and Dohl, T. P. (1980). The behavior of the Hawaiian spinner dolphin, *Stenella longirostris*. *Fish. Bull.* **77,** 321–349.

Norris, K. S., and Prescott, J. H. (1961). Observations on Pacific cetaceans of Californian and Mexican waters. *Univ. Calif., Berkeley, Publ. Zool.* **63,** 291–402.

Norris, K. S., Villa-Ramirez, B., Nichols, G., Würsig, B., and Miller, K. (1983). Lagoon entrance and other aggregations of gray whales, *Eschrichtius robustus*. *AAAS Sel. Symp.* **76,** 259–293.

Norris, K. S., Würsig, B., Wells, R. S., Würsig, M., Brownlee, S. M., Johnson, C., and Solow, J. (1990). "The Behavior of the Hawaiian Spinner Dolphin, *Stenella longirostris*." Univ. of California Press, Berkeley (in press).

Ohsumi, S. (1979). Interspecies relationships among some biological parameters in cetaceans and estimation of natural mortality coefficient of the southern hemisphere minke whale. *Rep. Int. Whal. Comm.* **29,** 397–406.

Ohsumi, S. (1986). Yearly change in age and body length at sexual maturity of a fin whale stock in the eastern North Pacific. *Sci. Rep. Whales Res. Inst.* **37,** 17–30.

Oliver, J. S., Slattery, P. N., Silberstein, M. A., and O'Connor, E. F. (1983). A comparison of gray whale, *Eschrichtius robustus*, feeding in the Bering Sea and Baja California. *Fish. Bull.* **81,** 513–522.

Payne, R. (1980). Research on the behavior of various species of whales. *Natl. Geogr. Soc. Res. Rep.* **12,** 551–564.

Payne, R., and Dorsey, E. M. (1983). Sexual dimorphism and aggressive use of callosities in right whales (*Eubalaena australis*). *AAAS Sel. Symp.* **76,** 295–329.

Payne, R., and McVay, S. (1971). Songs of humpback whales. *Science* **173,** 587–597.

Perrin, W. F., ed. (1986). "Newsletter of the Cetacean Specialist Group," No 2. IUCN-NMFS, La Jolla, California.

Perrin, W.F., and Reilly, S. B. (1984). Reproductive parameters of dolphins and small whales of the family Delphinidae. *Rep. Int. Whal. Comm., Spec. Issue* **6,** 97–133.

Pivorunas, A. (1979). The feeding mechanisms of baleen whales. *Am. Sci.* **67,** 432–440.

Reeves, R. R., and Katona, S. K. (1980). Extralimital records of white whales (*Delphinapterus leucas*) in eastern North American waters. *Can. Field Nat.* **94,** 239–247.

Reeves, R. R., and Mitchell, E. (1987). Cetaceans of Canada. *Underwater World* **59,** 1–27 (publ. by Dept. of Fisheries and Oceans, Ottawa, Ontario).

Reeves, R. R., Mitchell, E., Mansfield, A., and McLaughlin, M. (1983). Distribution and migration of the bowhead whale, *Balaena mysticetus*, in the eastern North American Arctic. *Arctic* **31,** 5–64.

Reeves, R. R., Ljungblad, D. K., and Clarke, J. T. (1984). Bowhead whales and acoustic seismic surveys in the Beaufort Sea. *Polar Rec.* **22,** 271–280.

Reilly, S. B. (1984). Observed and published rates of increase in gray whales, *Eschrichtius robustus*. *Rep. Int. Whal. Comm., Spec. Issue* **6,** 389–399.

Rice, D. W. (1974). Whales and whale research in the eastern North Pacific. *In* "The Whale Problem: A Status Report" (W. E. Schevill, ed.), pp. 170–195. Harvard Univ. Press, Cambridge, Massachusetts.

Rice, D. W. (1977). A list of the marine mammals of the world. *NOAA Tech. Rep.* **NMFS SSRF-711,** 1–15.

Rice, D. W., and Wolman, A. A. (1971). The life history and ecology of the gray whale (*Eschrichtius robustus*). *Am. Soc. Mammal., Spec. Bull.* **3,** 1–142.

Rice, D. W., and Wolman, A. A. (1981). "Summer Distribution and Numbers of Fin, Humpback, and Gray Whales in the Gulf of Alaska," Contract Rep. No. Ru-592. U.S. Minerals Manage. Serv., Reston, Virginia.

Richardson, W. J., Davis, R. A., Evans, C. R., Ljungblad, D. K., and Norton, P. (1987). Summer distribution of bowhead whales, *Balaena mysticetus*, relative to oil industry activities in the Canadian Beaufort Sea, 1980-1984. *Arctic* **40,** 93–104.

Rugh, D. J., and Fraker, M. A. (1981). Gray whale (*Eschrichtius robustus*) sightings in Eastern Beaufort Sea. *Arctic* **34**(2), 186–187.

Scammon, C. M. (1874). "The Marine Mammals of the North-Western Coast of North America, Described and Illustrated: Together with an Account of the American Whale-Fishery." J. H. Carmany & Co., San Francisco, California.

Schevill, W. E., Watkins, W. A., and Backus, R. H. (1964). The 20 cycle signals and *Balaenoptera* (fin whales). *In* "Marine Bioacoustics" (W. N. Tavolga, ed.), Vol. 1, pp. 147–152. Pergamon, New York.

Schmidly, D. J. (1981). "Marine Mammals of the Southeastern United States Coast and the Gulf of Mexico," Rep. U.S. Bur. Land Manage., U.S. Fish Wildl. Serv., Washington, D. C.

Scott, G. P., Burn, D. M., and Hansen, L. J. (1988). The dolphin die-off: Long-term effects and recovery of the population. *Proc. Oceans '88 Conf.*, Baltimore, pp. 819–823.

Sergeant, D. E. (1973). Biology of white whales (*Delphinapterus leucas*) in western Hudson Bay. *J. Fish. Res. Board Can.* **30**, 1065–1090.

Sergeant, D. E. (1977). Stocks of fin whales *Balaenoptera physalus* L. in the North Atlantic Ocean. *Rep. Int. Whal. Comm.* **27**, 460–473.

Sergeant, D. E. (1986). Present status of white whales, *Delphinapterus leucas*, in the St. Lawrence Estuary. *Nat. Can.* **113**, 61–81.

Sergeant, D. E., and Hoek, W. (1974). Seasonal distribution of bowhead and white whales in the eastern Beaufort Sea. *In* "The Coast and Shelf of the Beaufort Sea." (J. C. Reed and J. E. Sater, eds.), pp. 705–719. Arctic Inst. North Am., Arlington, Virginia.

Sergeant, D. E., Mansfield, A. W., and Beck, B. (1970). Inshore records of Cetacea of eastern Canada, 1949-68. *J. Fish. Res. Board Can.* **27**, 1903–1915.

Sergeant, D. E., Caldwell, D. K., and Caldwell, M. C. (1973). Age, growth and maturity of bottlenosed dolphins (*Tursiops truncatus*) from northeast Florida. *J. Fish. Res. Board Can.* **30**, 1009–1011.

Shane, S. H. (1980). Occurrence, movements and distribution of the bottlenose dolphin, *Tursiops truncatus*, in southern Texas. *Fish. Bull.* **78**, 593–601.

Shane, S. H., Wells, R. S., and Würsig, B. (1986). Ecology, behavior, and social organization of the bottlenose dolphin: A review. *Mar. Mammal Sci.* **2**, 34–63.

Silber, G. K. (1986). The relationship of social vocalizations to surface behavior and aggression in the Hawaiian humpback whale (*Megaptera novaeangliae*). *Can. J. Zool.* **64**, 2075–2080.

Silber, G. K. (1988). Recent sightings of the Gulf of California harbor porpoise, *Phocoena sinus*. *J. Mammal.* **69**, 430–433.

Slijper, E. J. (1979). *In* "Whales" (R. J. Harrison, ed.), 2nd ed. Hutchinson, London.

Smith, T. G., and Hammill, M. O. (1986). Population estimates of white whale, *Delphinapterus leucas*, in James Bay, Eastern Hudson Bay, and Ungava Bay. *Can. J. Fish. Aquat. Sci.* **43**, 1982–1987.

Smith, T. G., Siniff, D. B., Reichle, R., and Stone, S. (1981). Coordinated behavior of killer whales, *Orcinus orca*, hunting a crabeater seal, *Lobodon carcinophagus*. *Can. J. Zool.* **59**(6), 1185–1189.

Smith, T. G., Hammill, M. O., Burrage, D. J., and Sleno, G. A. (1985). Distribution and abundance of beluga, *Delphinapterus leucas*, and narwhals, *Monodon monoceros*, in the Canadian High Arctic. *Can. J. Fish. Aquat. Sci.* **42**, 676–684.

Tarpy, C. (1979). Killer whale attack. *Natl. Geogr.* **155**(4), 542–545.

Tavolga, M. C., and Essapian, F. S. (1957). The behavior of the bottlenosed dolphin (*Tursiops truncatus*): Mating, pregnancy, parturition, and mother-infant behavior. *Zoologica (N.Y.)* **42**, 11–31.

Tershy, B. R., and Breese, D. (1987). Body size, diet, habitat use and social behavior in *Balaenoptera* whales in the Gulf of California, Mexico. *Abstr., 7th Bienn. Conf. Biol. Mar. Mammal.*, Miami, Florida. (obtained from B. Tershy, Moss Landing Marine Labs, P.O. Box 450, Moss Landing, CA 95039).

Tershy, B. R., Breese, D., and Strong, C. (1990). "Abundance, Seasonal Distribution and Population Composition of Balaenopterid Whales in the Canal de Ballenas, Gulf of California, Mexico." *Rep. Int. Unol. Comm., Spec. Issue* 12 (in press).

Tyack, P. (1986). Population biology, social behavior, and communication in whales and dolphins. *Trends Ecol. Evol.* **1**, 144–150.

Vladykov, V. D. (1946). "Etudes sur les mammifères aquatiques. IV. Nourriture du marsouin blanc ou bluga (*Delphinapterus leucas*) du fleuve Saint-Laurent," Rep. Can. Dept. Fish. Oceans, Quebec.

Walker, W. A. (1981). "Geographical Variation in Morphology and Biology of Bottlenose Dolphins (*Tursiops*) in the Eastern North Pacific," NMFS Rep. No. LJ-81-03C. Natl. Mar. Fish. Serv., Southwest Fish. Cent., La Jolla, California.

Wells, R. S. (1986). Structural aspects of dolphin societies. Ph.D. Dissertation, University of California, Santa Cruz.

Wells, R. S., Irvine, A. B., and Scott, M. D. (1980). Social ecology of inshore odontocetes. *In* "Cetacean Behavior: Mechanisms and Functions" (L. M. Herman, ed.), pp. 263–317. Wiley (Interscience), New York.

Wells, R. S., Hansen, L. J., Boldridge, A., Dohl, T. P., Kelly, D. L., and Defran, R. H. (1990). Northward extension of the range of bottlenose dolphins along the California coast. *In* "The Bottlenose Dolphin" (S. Leatherwood and R. R. Reeves, eds.), pp. 421–431. Academic Press, San Diego.

Winn, H. E. (1984). Development of a right whale sighting network in the southeastern U.S. *U.S. NTIS PB Rep.* **PB 84-240548,** 1–12.

Winn, H. E., Goodale, D. R., Hyman, M. A. M., Kenney, R. D., Price, C. A., and Scott, G. P. (1981). Right whale sightings and the right whale minimum count. *In* "A Characterization of Marine Mammals and Turtles in the Mid- and North-Atlantic Areas of the U.S. Outer Continental Shelf," Cetacean and Turtle Assess. Program, pp. 49–51. University of Rhode Island, Kingston.

Würsig, B. (1978). Occurrence and group organization of Atlantic bottlenose porpoises (*Tursiops truncatus*) in an Argentine Bay. *Biol. Bull. (Woods Hole, Mass.)* **154,** 348–359.

Würsig, B. (1986). Delphinid foraging strategies. *In* "Dolphin Cognition" (J. Thomas, R. Schusterman, and F. Wood, eds.), pp. 347–359. Erlbaum, Hillsdale, New Jersey.

Würsig, B., and Würsig, M. (1979). The day and the night of the dolphin. *Nat. Hist.* **88,** 60–67.

Würsig, B., and Würsig, M. (1980). Behavior and ecology of dusky porpoises, *Lagenorhynchus obscurus*, in the south Atlantic. *Fish. Bull.* **77,** 871–890.

Würsig, B., Dorsey, E. M., Fraker, M. A., Payne, R. S., and Richardson, W. J. (1985). Behavior of bowhead whales, *Balaena mysticetus*, summering in the Beaufort Sea: A description. *Fish. Bull.* **83,** 357–377.

Würsig, B., Cipriano, F., and Würsig, M. (1990). Dolphin movement patterns: Information from radio and theodolite tracking studies. *In* "Dolphin Societies. Discoveries and Puzzles." (K. Pryor and K. Norris, eds.), Univ. of California Press, Berkeley (in press).

Zeh, J. E., Turet, P., Gentleman, R., and Raftery, A. E. (1988). Population size estimation for the bowhead whale, *Balaena mysticetus*, based on 1985 visual and acoustic data. *Rep. Int. Whal. Comm.* **38,** 349–360.

Zhou, K., Li, Y., Nishiwaki, M., and Kataoki, T. (1982). A brief report on the observations of the baiji, *Lipotes vexillifer*, in the lower reaches of the Yangtze River between Nanjing and Guichi. *Acta Theriol. Sinica* **2,** 253–255.

6

Physiologic and Toxic Effects on Cetaceans

Joseph R. Geraci
Department of Pathology
Ontario Veterinary College
University of Guelph
Guelph, Ontario
N1G 2W1 Canada

Historical Notes

January 28, 1969, marked the first incident calling attention to the possibility that oil pollution might harm a cetacean. Union Oil's undersea well off Santa Barbara, California, sprung a leak. A flurry of activity followed, with estimates and counterestimates of the size and duration of the spill and of the types and numbers of animals being affected. Union Oil calculated that about 250,000 gal of oil had leaked into the channel by February 8 when the well was sealed with cement (Santa Barbara *News-Press,* March 15, 1969). Oil continued to seep through new leaks and by April 1969, as much as 3 million gal may have been released (Santa Barbara *News-Press,* April 26, 1969; Straughan, 1972).

What happens to whales and dolphins when 800 square miles (Santa Barbara *News-Press,* February 6, 1969) of travel routes and feeding grounds are contaminated with oil? The picture is not clear. At the time of the spill, gray whales were just beginning their annual migration northward through or to the west of the fouled area (Brownell, 1971). One airplane pilot saw a group of whales moving northward through the slick, blowing as they went (Easton, 1972).

The Santa Barbara *News-Press* of March 13, 1969, reported that by the sixth week of the spill three dead gray whales had come ashore in northern California. The mouth and baleen plates of one were coated with

Sea Mammals and Oil: Confronting the Risks
Copyright © 1990 by Academic Press, Inc.
All rights of reproduction in any form reserved.

a light film of oil, later interpreted as an unremarkable finding in a carcass that had floated at sea for some time (Brownell, 1971). Another whale covered with oil was seen "floating listlessly" in the Santa Barbara Channel. By March 15 the number of dead whales was placed at six (Santa Barbara *News-Press,* March 15, 1969), only two of which were fresh enough for necropsy examination. Oil was not found in either, and the whales were thought to have died from natural causes (Santa Barbara *News-Press,* May 15, 1969). One, in fact, may have been harpooned (Brownell, 1971).

During the Union Oil spill, gray whales were the focus of attention for their obvious size, and possibly because they were protected after a long history of exploitation (Rice and Wolman, 1971). But they were not the only cetacean to arouse concern. An unidentified porpoise "coated with old oil" stranded on a sandbar. Two other stranded porpoises examined by representatives of the California State Department of Fish and Game showed no evidence that oil was related to their deaths (Battelle Memorial Institute, 1969). Eyewitnesses, referring to one of the porpoises, a common dolphin (Brownell, 1971), said that its "breathing hole was clogged with oil and its lungs were ruptured" (Easton, 1972). *Time* magazine (February 21, 1969) published a similar account. Yet Brownell (1971) and his colleagues examined five of six dolphins stranded from February through May 1969 and found no evidence of oil contamination.

The final tally? Carcasses recovered from January 28 through March 31, 1969, included six gray whales, one sperm whale, one pilot whale, five common dolphins, one Pacific white-sided dolphin, and two unidentified dolphins. Brownell (1971), acknowledging that these totaled more than the usual number of gray whales and dolphins stranding annually on California shores, concluded that increased survey effort in 1969 had led to higher counts. The Smithsonian Institution Center for Short-lived Phenomena (Anonymous, 1970) summarized the event: "a few sea mammals were found dead; however, for the most part, they seemed to avoid direct contact with the oil." An independent report went a step further in concluding that "the whales were either able to avoid the oil, or were unaffected when in contact with it" (Battelle Memorial Institute, 1969).

Not surprisingly, the Santa Barbara spill set the stage for public reaction toward the threat of oil to whales and dolphins. The oil flowed for a period of months in full view of a newly aroused public. Until that time little thought had been given as to why an occasional whale or dolphin might come ashore. Now, the need for such answers had become crucial, and with neither answers nor an established protocol to determine how a pollutant might fit into the picture, the stage was set for futile speculation.

Vying in magnitude with the Santa Barbara incident was the

grounding of the *Exxon Valdez* in Prince William Sound, Alaska, in March, 1989. Eleven million gallons of Prudhoe Bay crude oil were released, causing a flurry of activity. There was concern that, of the 15 or 20 species of cetaceans that occupy the area periodically, some might become victims of the spill. By October, 1989, biologists had documented the carcasses of 25 gray, 2 minke, 1 fin, and 3 unidentified whales, and 6 harbor porpoises. At the time of this writing, the investigation into the cause of their deaths is underway. The tally of dead gray whales was higher than previously recorded, yet may not be extraordinary considering virtually the entire population of 21,000 gray whales had migrated through the area. They would normally leave in their wake a number of carcasses, which in most years would go unobserved along remote stretches of beach. It remains to be seen how many of the animals were actual victims of the spill.

Most of the other incidents have been minor by comparison (Table 6–1). Following a spill of light diesel fuel along the Alaskan shore, two killer whales, one sick and one dead, were reported. There was no additional detail (Anonymous, 1971). A ruptured storage tank released more than 11 million gal of hot Bunker C oil into Japan's Inland Sea. A press report revealed that one porpoise had died (Nicol, 1976). After the *Amoco Cadiz* ran aground spilling 60 million gal of light crude oil along the coast of the Brittany region of France, six badly decomposed cetacean carcasses were examined for evidence of oil. Prieur and Hussenot (1978) noted that one may have shown signs of oil contamination, while the remainder were species which commonly strand. The observers concluded that any relationship between the oil spill and the death of the animals would have been difficult to establish. The *Hellenic Carrier* collided with a ship on the outer banks off Nags Head, North Carolina, and sank, spilling 3000 gal of oil. A report (Anonymous, 1981) told of a porpoise that was killed.

On March 21, 1982, the decomposed carcass of a male pilot whale was found stranded in Rodanthe, North Carolina. On its skin was a 10 × 20-cm patch of dry tar (Anonymous, 1982). No detail was provided. During a survey of cetaceans in the western North Atlantic, a dead *Grampus* was spotted a few kilometers away from an extensive oil sheen (Sorensen *et al.*, 1984).

What conclusions can be drawn from these observations? It seems that unlike sea otters, polar bears, and some seals, there is no gripping evidence that oil contamination has been responsible for the death of a cetacean.

Reactions of Cetaceans to Oil

The Battelle Memorial Institute's (1969) summary of the Santa Barbara spill concluded that whales may have been able to avoid the oil. That casual

Table 6–1
Reports of Cetaceans Associated with Oil

Date	Location and source	Oil type and quantity	Species	Impact	Reference
Feb. 1969	Santa Barbara, Calif.; Union Oil well	Crude oil, $>30 \times 10^6$ gal	Gray, pilot, and sperm whales; common and white-sided dolphins	16 stranded whales and dolphins recovered. No causal relationship	Brownell (1971)
Apr. 1970	Alaska Peninsula	Diesel fuel, quantity ?	Killer whale	1 sick and 1 dead animal observed. No examination.	Anonymous (1970)
1974	Japan	Bunker C, 11.3×10^6 gal	Porpoise	1 dead porpoise found.	Nicol (1976)
Oct. 1976	Aransas Pass, Texas; pipeline leak	Crude oil, 15,500 gal	Bottlenose dolphin	Dolphins swam through oil without apparent effect.	Shane and Schmidly (1978)
Dec. 1976	Nantucket Shoals; *Argo Merchant*	Bunker C, 7.9×10^6 gal	Fin whale, pilot whale, others	43 sightings of animals in and around patches of oil. No obvious reaction.	Grose and Mattson (1977)
Mar. 1978	France; *Amoco Cadiz*	Crude oil, 60×10^6 gal	White-sided and common dolphins; pilot whale	6 stranded animals with no firm evidence of oil.	Prieur and Hussenot (1978)
Sept. 1978	Matagorda Bay, Texas; boat grounding	Fuel oil, 3000 gal	Bottlenose dolphin	20 dolphins swimming through oil without effect.	Gruber (1981)

June 1979	Gulf of Mexico; Ixtoc-I	Crude oil, 70×10^6 gal	Bottlenose and spotted dolphins	Animals sighted in areas with oil-coated debris. Apparently unaffected.	Bergey (1979)
June 1979	Cape Cod, Mass.; *Regal Sword*	Bunker C, 80,000 gal Fuel oil, 6300 gal	Humpback, fin, minke, right whales; white-sided dolphins	Animals feeding, surfacing and swimming through heavy concentrations of oil.	Goodale *et al.* (1981)
May 1981	Outer Banks, N. Car.; *Hellenic Carrier*	Type ?, 3000 gal	Porpoise	Unconfirmed report of dead porpoise.	Anonymous (1981)
Mar. 1982	Rodanthe, N. Car.; Source ?	Tar	Pilot whale	Stranded whale with small patch of dry tar on skin.	Anonymous (1982)
July 1984	Gulf of Mexico; *Alvenas*	Crude oil, $>1 \times 10^6$ gal	Bottlenose dolphin	1 dolphin swimming in the midst of oil patches. Others at the edge of the slick.	Owen (1984)
Mar. 1989	Prince William Sound, Alaska; *Exxon Valdez*	Crude oil, 11×10^6 gal	25 gray, 1 fin, 2 minke, and 3 unidentified whales; 7 harbor porpoises	Stranded carcasses. Possible unrelated natural mortality.	H.W. Braham (personal communication)

[a] All volumes converted to U.S. gallons, and rounded. 1 barrel = 159 L = 42 gal. 1 ton = 278 gal of bunker oil, 300 gal of crude oil, or 332 gal of diesel fuel.

observation, it now appears, may have some empirical footing. Fragmentary data from subsequent spills support the notion that whereas some cetaceans may avoid oil, others, willing or not, enter it without obvious peril. Shane and Schmidly (1978) studied a group of *Tursiops* in Aransas Pass, a few kilometers north of Corpus Christi, Texas (Fig. 6–1). On October 14, 1976, 15,500 gal of crude oil leaked from pipelines along Aransas Channel and into Morris and Cummings Cut. "The dolphins swam regularly through the oil slick, although they were not observed surfacing in the heaviest concentrations of oil. [They] began feeding and mating once they reached cleaner water. The oil had no obvious effect upon them" (Shane and Schmidly, 1978). That incident was soon paralleled by an event which took place in Matagorda Bay, 100 km north of Aransas Pass (Fig. 6–1). A tugboat ran aground spilling some 3000 gallons of fuel oil into Pass Cavallo where Gruber (1981) was studying the behavior

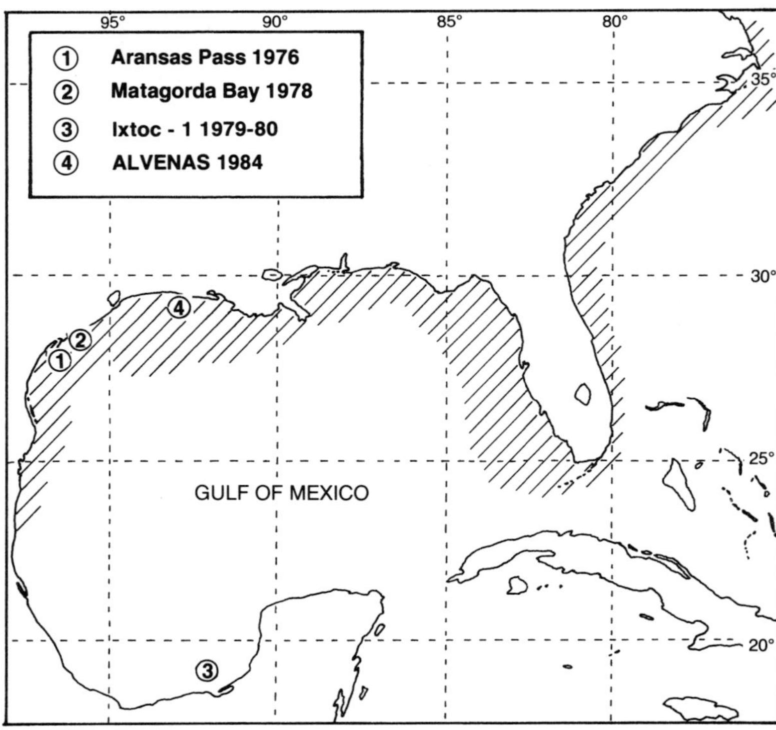

Figure 6-1
Cetaceans have been observed in the vicinity of at least four oil spills in the Gulf of Mexico.

of *Tursiops*. She noted that 20 dolphins, including one calf, would swim back and forth through "large globules" of oil in an extremely polluted section of the intracoastal waterway. The animals repeatedly surfaced in the midst of the thickest concentrations, while playing and tossing fish to one another. Most of the dolphins were precisely in the areas where the slicks were abundant.

Farther north, events were unfolding on a larger scale. On December 15, 1976, the *Argo Merchant* ran aground off Nantucket Island in Massachusetts, spilling nearly 8 million gal of No. 6 fuel oil. Between December 28, 1976, and January 13, 1977, aerial observers recorded 43 separate sightings of cetaceans, including 21 fin, 2 other unidentified rorquals, 7 unidentified dolphins, and 13 to 15 pilot whales. On one occasion, two finback whales were in a heavily oiled area, showing no apparent reaction to it. Limited data showed that there was no bias in the animals' distribution in relation to oil. No marine mammal was seen in obvious distress or in direct physical contact with oil pancakes or sheen (Grose and Mattson, 1977).

A University of Rhode Island research group, whose project was known by the acronym CETAP (Cetacean and Turtle Assessment Program), conducted a 3-year systematic survey of a 210,000 km^2 area of the western North Atlantic. They spotted oil 94 times, and cetaceans were twice seen within oil. Both sightings were of pods of common dolphins. They noted no behavior other than swimming (Sorensen *et al.*, 1984). On June 18, 1979, the CETAP team was on scene to investigate the aftermath of a collision between two tankers southeast of Cape Cod, Massachusetts. One, the *Regal Sword*, sank, liberating 86,300 gal of Bunker C and No. 2 fuel oil. Over the next week, at least three and possibly four species of cetaceans were seen within the slicks. Humpback and fin whales were observed feeding at the surface, some in the middle of a heavy slick. One whale, tentatively identified as a right whale, repeatedly surfaced in oil. Whales and a large number of white-sided dolphins swam, played, and fed in and near the slick. Dolphins were seen in oil slicks and sheens during 8 of the 10 flight passes made over the most heavily polluted areas. The investigators reported that there was no difference in behavior between cetaceans within the slick and those beyond it (Goodale *et al.*, 1981).

The tanker *Alvenas* ran aground and ruptured off the coast of Louisiana (see Fig. 6–1), spilling more than a million gallons of crude oil (Owen, 1984). Aerial observers found heavy slicks extending 2 miles offshore. One *Tursiops* was sighted amid patches of oil inshore; several others were offshore at the outer edge of the slick. There were no dolphins in the immediate vicinity of the vessel.

It is ironic that the search for marine mammals surrounding the largest recorded oil spill was, by any standard, modest. Between June 3,

1979, and March 23, 1980, the Ixtoc-I oil well in the Bay of Campeche in the Gulf of Mexico (see Fig. 6–1) had a blowout which leaked a million gallons of oil daily. Spinner dolphins, bottlenose dolphins, and unidentified porpoises were observed in areas containing oil-coated debris; all appeared healthy and free of oil. Porpoises that were bow-riding veered to avoid tar balls, their only obvious reaction to oil (Bergey, 1979).

Evans (1982) from Hubbs-Seaworld Research Institute led a team of investigators to study the reaction of migrating gray whales to natural oil seeps emanating from the sea floor. Thus, our story begins and ends with the gray whale. The team established four observatories on land and one on an offshore drilling rig, along a 50-km stretch of the California coast from Point Conception to Coal Oil Point. Within the study area, there were at least four seeps within a 5-km radius of each observation site. The most active seep, near Coal Oil Point, releases a minimum of 30 barrels of oil daily.

From the observatories and from aircraft, the investigators documented swimming speeds, surface behavior, dive times, and respiratory rates of small groups of whales, and found that when entering oiled waters, the animals would modify their swimming speeds and occasionally their direction with no consistent pattern.

In oiled waters they seemed to spend less time at the surface, blowing less frequently but faster. If this reaction is interpreted as an avoidance response, it suggests that gray whales can detect oil. Those showing no response either could not detect the amount or type of oil present or were indifferent to it. The investigators were careful to point out that comparisons were tenuous, as it was not possible to follow specific whales into and out of oiled areas. That should be the subject of future experiments and could well be incorporated into ongoing studies on the behavior of radio-tagged gray whales during their annual migration.

These observations summarize the attempts to determine what cetaceans do when they confront spilled oil. None provides a complete picture, as one might expect from empirical studies. But each account gives a clue to behaviors, some more consistent than others, on which hypotheses can be constructed and tested in subsequent oil spill accidents. The denouement of certain mortality no longer seems reasonable. Instead, we find whales and dolphins in the vicinity, and some in the midst of a spill, behaving quite normally. The questions now are: Were the animals able to detect oil; given the choice, would they have avoided it; might they have been drawn unwittingly to the heart of a spill, perhaps in search of prey organisms attracted by the oil's protective shadow; how might such excursions through oil affect the health of a whale or dolphin? These questions have been addressed through an assortment of speculative writings (Sergeant, 1970; Butler *et al.*, 1974; Fraker *et al.*, 1978; Calkins, 1979; Geraci and St. Aubin, 1980; Cowles *et al.*, 1981) and, more recently, by a series of experiments conducted by the author's research team.

Detection and Avoidance

From 1980 through 1983, we carried out three successive studies to determine how bottlenose dolphins react to oil films in their environment (Geraci *et al.*, 1983; Smith *et al.*, 1983; St. Aubin *et al.*, 1985). The first of these tested the hypothesis that surface oil presents a visual target to an animal. Most crude oils are dark, and as they weather, become thicker and darker still. We intended to learn whether a bottlenose dolphin would be able to detect oils of this kind and dilute preparations with less apparent visual properties. In all, two dolphins were presented with 12 different oils and 22 mixtures.

The studies were carried out in the relatively natural setting of an enclosed lagoon in the Gulf of Mexico in the Florida Keys. Early into the training, each dolphin learned to position itself on a fixed underwater station at a depth of 1 m, and look upward to view a short open-ended cylinder confining various materials and objects at the surface (Fig.6–2).

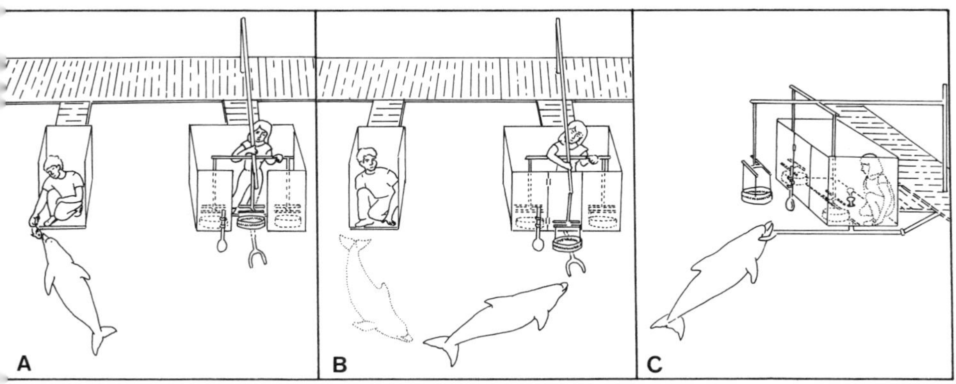

Figure 6-2
Testing procedure to determine whether bottlenose dolphins can detect oil. (A) The trainer calls the animal to the resting dock while an examiner prepares up to three test cylinders in a canal concealed within the trial dock. (B) On command, the dolphin swims to the trial dock, stations on a stirrup, and (C) examines the contents of the cylinder. It leaves the station to press the paddle only when it detects there is something (oil) in the cylinder besides water. The study demonstrated that dolphins quite easily discriminate between oil and the uncontaminated surface of the water (Geraci *et al.*, 1983).

The animal would leave the station to press a paddle only when it detected something in the cylinder besides water. If the dolphin correctly detected nothing, it would remain stationary until called to be rewarded. The behavior was shaped using solid objects made of wood and plastic, and an assortment of buoyant fruits, selected according to the palate of each trainer. After the animal mastered the technique, the objects were gradually replaced with oil, until the cylinder contained only oil. Once trained to detect solid objects, it took only three days for the dolphins to respond to the presence of oil alone.

Through a long series of randomized presentations of different objects and types and thicknesses of oil, the experiment demonstrated that the dolphins quite easily discriminated between oil and the uncontaminated surface of the water. The darker the substance, the easier it was to detect, down to an optical density that corresponded to 1-mm thick films of the three types of crude oils tested. We interpreted these findings as evidence that the dolphins can see the thicker formations of oil that typically occur at the source of a spill, and also weathered fractions which form "pancakes" of much thicker viscous oils. In fact, while blindfolded, one of the dolphins was able to detect presumably by echolocating, 12-mm-thick patches of heavy oil churned so as to entrain air bubbles. Lighter fractions, which spread into thin sheens and typically comprise most of a spill area, would not be detected easily, if at all, nor could lightly colored refined products such as gasoline, diesel fuel, and solvents which disperse very rapidly into surface films. Patterns of weathering and spread of oil are covered in detail by Neff (this volume, Chapter 1).

The study answered a fundamental question and, quite naturally, raised several others. If dolphins can see oil, why do they enter the region of a spill? Might they have a compelling reason to be there? Perhaps they do reject the silhouettes of thicker oil and penetrate only the less visible sheens. This, of course, could bring them into contact with volatile, more acutely harmful substances. These concerns were incorporated into the design of the next experiment.

The objective was to determine if bottlenose dolphins would avoid a detectable slick. To begin with, three dolphins that, as far as we knew, had not been exposed to oil were allowed to roam freely in an oceanic pen, with the surface divided into three equal areas by oil-containment booms (Fig. 6–3). Observations were made while the animals were in the pens, either alone or as a group. We thereby established their individual swimming and surfacing behaviors and their "desire," if any, to occupy one of the three subdivisions. Following the observation period, the dolphins were removed while an oil slick 1 cm thick was added to one of the subdivisions, then reintroduced individually or as a group. To avoid harm-

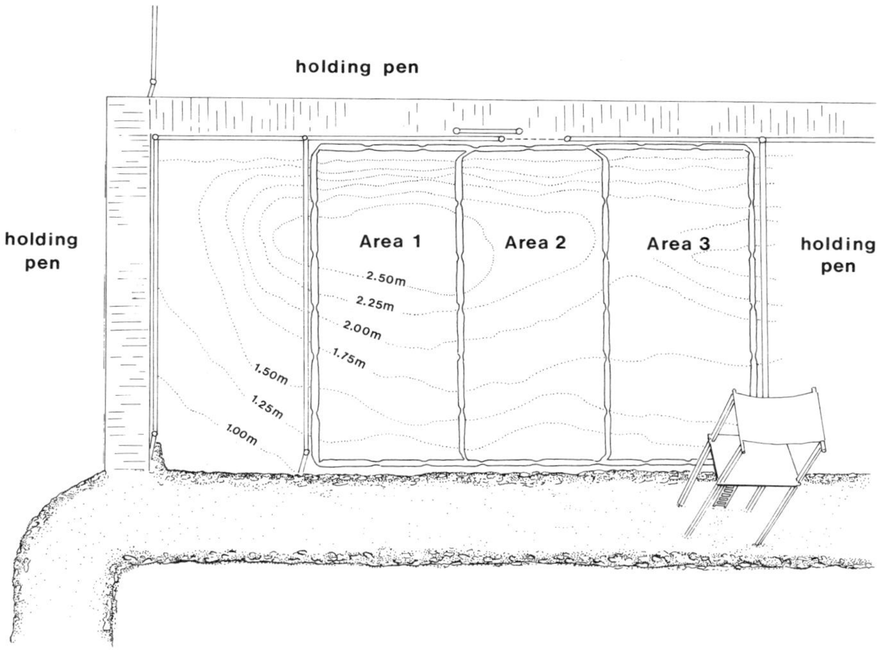

Figure 6-3
Test setting to determine a dolphin's reaction to an oil-covered sector (1, 2, or 3). Each of three dolphins, introduced into the setting, avoided the oiled subdivision for up to 52 minutes. Then each emerged in the oil once on the first day, and never again for the duration of the study (Smith *et al.*, 1983).

ing the dolphins or complicating the experiment, only odorless, tasteless, nontoxic, highly refined mineral oil was used. It was mixed with a black colorant to match the optical density of crude oils that dolphins had been able to detect in the previous study.

After re-entering the pen individually, all three dolphins avoided the oiled subdivision for at least 5 minutes, and one for up to 52 minutes. This initial reluctance was regarded as a probable response to any new stimulus, and not to oil *per se*. Within an hour, however, each dolphin emerged in the oil either accidentally, or as part of an investigative process. Each reacted immediately and overtly with a "startle" response and behavior normally associated with stress or annoyance. Yet the oil was innocuous, indicating that another physical property of the oil, viscosity perhaps, had been enough to disturb them. The behavior suggested that tactile sense may have played a role in the dolphins' reaction to oil.

After the initial contact, the dolphins never again emerged in oil, even when reintroduced four days later to the experimental setting. In fact, their aversion to it prevented them from swimming beneath the oil to adjacent uncontaminated pens. The dolphins had developed an aversion to oil, not unexpected perhaps of a creature genetically driven to regard the water's surface as a secure portal to clean air. But what surprised us was the apparent ability of the animals to "feel the oil."

We undertook a third study, designed to show how sensitive was the tactile response, and how thin or clear an oil slick would have to be for a dolphin to disregard it. We used the same experimental setting, but this time tested the dolphins' response to oils we thought would be less obvious. We kept testing their reaction to the dark-colored mineral oil, more or less as a control, to see whether they might eventually become accustomed to it. They apparently did not because they consistently avoided it. We also presented two of the dolphins with the same mineral oil without the colorant. Both contacted it a total of four times within 15 minutes of exposure, each time showing a marked startle response. They never again touched the clear oil on that or the two subsequent trial days. We established a sheen of 0.1-mm nominal thickness, using automotive motor oil. In that setting, their behavior was erratic. A dolphin would surface there 100 times in a 1-hour session, and not at all in another, perhaps due to the inevitable discontinuities of such a thin sheen, or to reduced cutaneous stimulus presented by the film.

We repeated the study under vastly subdued light to reduce some of the visual properties of the oiled surface. This was done by covering the study pools with a large tent made of polypropylene shade fabric designed to screen out 92% of incoming light (Fig. 6–4). At night there was now insufficient light to activate a conventional light meter. In the pitch-black setting, while observers were stumbling about, dolphins were nimbly avoiding both the colored and clear mineral oil preparations. It appeared that the colorant had been incidental to the studies; the principal cue was cutaneous detection of oil. Their reaction to oil sheens under these conditions was the same as in daylight. One dolphin avoided the area containing the sheen; the other two swam into it, but less frequently.

It has become clear, through these studies, that bottlenose dolphins are able to detect and avoid a variety of oils both during the day and at night. To accomplish this, they rely predominantly on vision, and to some extent echolocation, when facing thick transparent slicks. Once a dolphin surfaces in oil, irrespective of light conditions, its response thereafter is to avoid it. That behavior, it seems, is triggered or reinforced by the sensation that oil creates on the animal's skin.

It is not certain how broadly the findings from the bottlenose

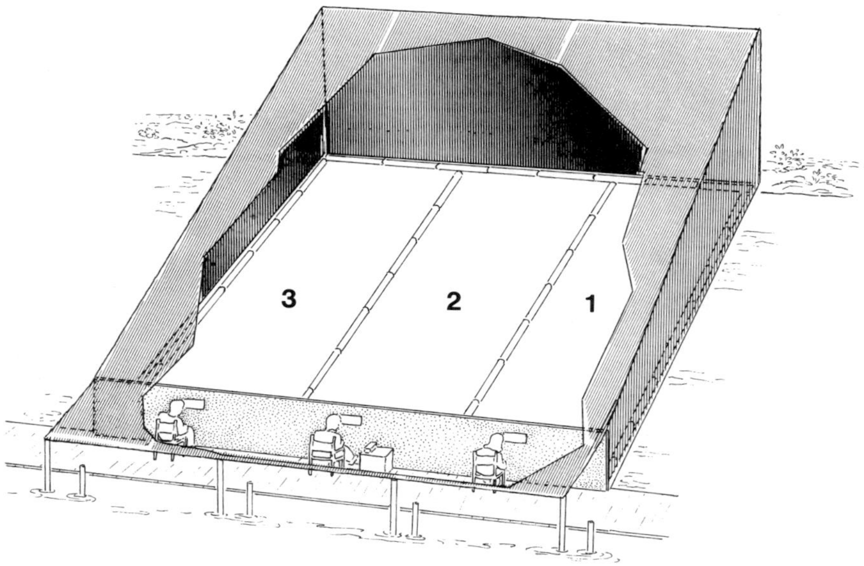

Figure 6-4
Test pool to determine, under subdued light, a dolphin's reaction to an oil-covered sector (1, 2, or 3). The shade canopy screened out 92% of moon- and starlight. While observers were stumbling about, dolphins nimbly avoided both colored and clear mineral oil preparations (St. Aubin *et al.*, 1985).

dolphins relate to other species of cetaceans. A case can be made for other dolphins and toothed whales with comparable sensory capabilities. Using analogies cautiously, certain of these data may be applied to mysticete whales as well. The key element is whether an animal has the sensory capacity to detect oil as the dolphins did, using vision and touch. Studies show that odontocetes have effective underwater vision, and aerial vision comparable to that of many terrestrial carnivores (Spong and White, 1971; Pepper and Simmons, 1973; Dawson, 1980; Madsen and Herman, 1980).

The mysticete eye is proportionately smaller and may not function as effectively as that of an odontocete at comparable depths (Mann, 1946; Walls, 1963; Waller, 1984). However, at or near the surface, their eyesight seems to be quite good. Indeed, mysticete whales may rely on visual cues for orientation. "Scouting" and "spy-hopping" is a typical behavior in which a whale may rear or bob out of the water, apparently viewing surface features and the shoreline (Cummings and Thompson, 1971; Eberhardt and Evans, 1962; Pike, 1962). Bowhead whales have been

A southern right whale pokes its head above the water, perhaps to obtain visual cues for orientation. Photo by B. Würsig.

seen preoccupied with floating logs and with bright green sheens of fluorescent dye (Würsig *et al.,* 1982). Echelon swimming, in which whales move in unison, appears also to rely on visual signals, as does the ability of humpbacks to pace whale-watching vessels, and of bowheads to punch holes selectively in newly forming ice (Ljungblad *et al.,* 1983). Such evidence indicates that mysticetes have evolved with odontocetes to rely on vision for spatial orientation and navigation to some extent. Should oil be present in a form which sufficiently alters the optical properties of the surface, a variety of cetacean species seems to have the visual capability of detecting it.

There are many lines of evidence to show that cetaceans generally have well-developed cutaneous sensitivity. Odontocete skin has nerve endings which in other mammals function in part as mechano-receptors (Palmer and Weddell, 1964; Schmidt, 1977; Agarkov *et al.,* 1974; Slijper, 1979; Purves and Pilleri, 1983). The structures are found particularly in the head region and in the skin of the lips, rostrum, and melon. It would be useful to know whether these nerve endings are responsible for receiving the cutaneous signals that reinforced the oil-avoidance behavior by the dolphins. If so, it might be possible to predict more accurately the probable response to oil by a particular cetacean with similarly endowed skin (Herman and Tavolga, 1980). Receptor-like structures have also been observed in the integument of mysticetes, including the fin whale (Giacometti, 1967). Virtually the entire surface of the mouth of baleen whales contains modified Golgi-Mazzoni corpuscles considered by Ogawa and Shida (1950) to be highly tactile organs. Vibrissae along the snouts of mysticetes (Yablokov *et al.*, 1974) and non-myelinated nerve fibers within dermal papillae of bowhead whales (Haldiman *et al.*, 1985) are also thought to be related to epidermal sensation.

Hence, it seems that the skin of the great whales is suitably equipped to receive cutaneous signals. It is not surprising in view of their obvious responsiveness to touch. In fact, this sensory mode may underlie basic affiliative and courtship behaviors (Herman and Tavolga, 1980), and for certain species like the bowhead whale, a sensitive tactile response would be a useful aid when navigating through vast fields of ice. Research into the nature and sensitivity of these and other sensory receptors would ignite a new level of thinking while providing fundamental data needed to extrapolate to species which cannot be tested experimentally.

Surface Contact

We assume that large whales and dolphins must be affected by contacting crude oil or a petroleum compound. But how? Because they have no fur, we are not concerned over loss of insulation. Instead we suggested that cetacean skin might, because of its unusual properties, respond to noxious substances in a manner approaching sensitive mucous membranes, with consequent effects on ionic regulation and water balance (Geraci and St. Aubin, 1980). The literature on humans accidentally contacting oil products provides some clues as to how studies on cetaceans might be developed.

Petroleum compounds, especially the short-chain fractions in gasoline, typically irritate skin and mucous membranes (Dutton, 1934; Hansbrough *et al.*, 1985). This irritation is due in part to solubilizing and removing cutaneous lipids (Wolfram *et al.*, 1972; Cornish, 1980), triggering an inflammatory response which first appears as reddening of the skin (Hansbrough *et al.*, 1985). Persistent contact causes necrosis (Walsh *et al.*, 1974) and inflammation—reactions which can be mapped and described quantitatively.

With that background, we designed a number of experiments to test how exposure to petroleum hydrocarbons might damage cetacean skin (Geraci and St. Aubin, 1982, 1985). We took a cautious approach, which began by applying a small sponge soaked in crude oil to discrete areas of skin of four species of odontocetes. Contact for up to 45 minutes was ineffective, in marked contrast to similar tests on human volunteers. We then progressed to longer exposures, up to 75 minutes, with gasoline. At this point it became clear that even unrealistically long contact times could not elicit the kind of severe reaction that typically occurs in other mammals (Hunter, 1968; Hansbrough *et al.,* 1985). Subtle changes that did occur were evident only histologically and, in each case, healed within a week.

The studies pointed to the effectiveness of cetacean epidermis as a barrier to the noxious substances found in petroleum. Whereas these normally damage the skin by permeating intercellular spaces and dissolving protective lipids, their penetration in cetacean skin was impeded by tight intercellular bridges, the vitality of the superficial cells, and the extraordinary thickness of the epidermis. The intercellular and intracellular lipids which are abundant in cetacean epidermal cells, and which we had assumed to be a vulnerable target for petroleum, were unaffected. In fact, they are protected well enough that after exposing skin from a white-sided dolphin to gasoline for 16 hours *in vitro*, we could not detect a change in lipid concentration (Geraci and St. Aubin, 1985).

By then, we had completed a study on repair of superficial wounds in the skin of *Tursiops*. An important finding was that following a cut, newly exposed epidermal cells degenerate to form a zone of dead tissues which shields the underlying cells from seawater during healing. We wanted to determine how oil might affect this process. For 30 minutes we massaged cuts with crude oil or tar. The substances had no effect on healing. Applied in the same manner, leadfree gasoline caused an exaggerated inflammatory response, which by 24 hours subsided and was indistinguishable from control cuts. We concluded that the devitalized shield had protected underlying tissue from gasoline in the same way it repels osmotic attack by seawater.

Biochemical and metabolic probing did reveal subtle reversible changes in cells exposed to petroleum. Each of these effects could have been explored in greater depth, but the exercise would not have provided a clearer understanding of the issue. Already the studies had progressed beyond a probable scenario for oil contact at sea. A script can be created in which a dolphin or whale is trapped in fresh oil rich with volatile short-chain fractions which are toxic when inhaled or ingested. Effects on mucous membranes would be inevitable, but of lesser concern. Spilled crude oil exists in this form only briefly. A cetacean is more likely to contact weathered oil, which is far more persistent but contains little of the more toxic light hydrocarbon fractions. Studies show that, in real life, contact with oil would be less harmful than we and others had proposed (Geraci and St. Aubin, 1980; Albert, 1981).

Inhalation

To some, oil spills conjure an image of a sea blackened with a thick coagulum which can endanger a whale by clinging to its surface and preventing it from feeding. The scene may reflect our own experience with tar ball-strewn beaches, but a greater threat to whales or dolphins is not the thick murky residue, but the invisible gaseous compounds that escaped from it. Vapors arise from volatile fractions in fresh crude oils and many of the refined products (Neff, this volume, Chapter 1). They irritate and damage soft tissues such as mucous membranes of the eyes (Carpenter *et al.,* 1977) and airways. Depending on the concentration of vapors and duration of exposure, their effects range from mild irritation (Valpey *et al.,* 1978) to sudden death (Wang and Irons, 1961).

On a positive note, vapors dissipate rapidly from the environment. Few investigators have analyzed vapor concentration or characteristics associated with a spill, perhaps because there has been little concern regarding their effect on humans. A cetacean, however, must draw its breath from the narrow blanket of air immediately overlying the surface of oil (or water), thereby intensifying its exposure. What will it inhale? In an early study (Geraci and St. Aubin, 1982), we calculated the concentrations of hydrocarbons associated with a theoretical spill of a typical light crude oil. We made the improbable assumption that all of the volatile substances in a 5-mm slick would evaporate instantaneously and completely into a 1-m layer of static air above the surface, thereby exposing an animal to an artificially maximized concentration of vapors. For each volatile compound, we calculated vapor pressure and concentration, then graphed the findings with those from toxicity studies on experimental animals other than whales.

The results showed that vapor concentrations could reach critical levels for the first few hours after a spill. A whale or dolphin unable to leave the scene during that time would inhale vapors and might be harmed. For a given exposure, the effect would depend on the health of the animal and its immediate response to stress (Thomson and Geraci, 1986). A panicking whale or swiftly moving dolphin would breathe rapidly and probably inhale more vapors. If this behavior were aggravated by excessive release of adrenalin, sudden mortality could result, as has been observed occasionally in humans (Bass, 1986). More likely, the animals would experience some irritation of respiratory membranes and absorb hydrocarbons into the bloodstream, a process which might be facilitated by their habit of submerging with full lungs. Whatever the mechanism, it is clear that for the short time they persist, vapors are one feature of an oil spill that can threaten the health of a cetacean.

Baleen Fouling

A great deal of interest has been expressed in the possibility that residues of oil may adhere to baleen plates so as to block the flow of water and interfere with feeding. The concerns are largely speculative (Fraker *et al.*, 1978; Calkins, 1979; Albert, 1981; Fritts *et al.*, 1983; Hansen, 1985), as there is only one relevant report, that of a gray whale found dead during the Santa Barbara oil spill with a light film of oil and dirt on its baleen plates (Brownell, 1971). Such an effect might be imperceptible, though leading to subtle, long-term consequences to the affected animal. With that in mind, two independent studies have been undertaken on the effects of oil fouling on baleen whales.

The feeding apparatus consists of two rows of fringed horny plates set into the gum tissue of the upper jaw. The plates are formed of hair-like tubules embedded in tough flexible keratin. The tongue of the animal rubs against the inner margin of the plates, abrading them and exposing the hairs which entwine to form a dense sieve. After the animal has taken a mouthful of food, pressure of the tongue against the plates drives water through the sieve, leaving behind the mass of food which the tongue delivers to the esophagus.

Baleen structure varies with the feeding habits of the whale. Right whales typically skim the surface, while rorquals (eg., fin, blue, and humpback whales) gulp their food. Gray whales are unlike other mysticetes in scouring the bottom in search of infaunal benthic species which they gather along with silt, sand, and gravel (Rice and Wolman, 1971). Pivorunas (1976, 1979) has written a detailed account of the relationship between baleen structure and feeding habits, and Würsig (this volume, Chapter 5) has examined the flexibility of feeding strategies within each group.

A safe assumption is that any substance in seawater which alters the characteristics of the plates, the integrity of the hairs, or the porosity of the sieve may jeopardize the nutritional well-being of the animal. A series of studies was conducted to determine whether petroleum compounds were capable of such mischief (Braithwaite, 1981; Geraci and St. Aubin, 1982, 1985; St. Aubin *et al.*, 1984).

Our studies began by evaluating the effects of various petroleum hydrocarbons on isolated baleen plates. Samples from seven species of whales were soaked in gasoline, crude oil, or tar, some for unrealistically long periods of time so as to exaggerate changes which might otherwise have been difficult to detect. For example, plates were exposed to crude oil for 8 hours, gasoline for up to 14 days, and roofing tar (our commercially available equivalent of weathered oil) for 21 days. Subsequently, the plates were tested for their breaking strength by tensiometer, analyzed for keratin integrity by x-ray diffraction, and finally, ground, ashed, and subjected to elemental analysis by atomic absorption spectrophotometry and colorimetry.

Immersion in gasoline for 90 minutes or in crude oil for 8 hours had no effect on protein structure or breaking strength of the plates. After 21 days in tar, x-ray diffraction patterns showed no change related to protein degradation (St. Aubin *et al.*, 1984). Nitrogen concentrations increased in all immersed samples, likely resulting from the loss of lipids which normally comprise up to 10% of the dry weight of baleen. There was also a consistent decrease in concentrations of manganese, copper, boron, and iron in exposed baleen hairs, but not plates, of fin and gray whales. Right whale samples were unaffected. There was no tensiometric evidence of increased plate fragility associated with these changes in elemental composition. We concluded the study at this point and directed our emphasis toward determining the effects of oil on baleen function.

There is very little information on how the baleen apparatus actually operates, which complicates the design of any experiment to determine how oil fouling affects function. The main difficulties in our study were in estimating water flow rates and pressure across the baleen filter, impediments to flow under normal conditions of feeding, and the functional reserve capacity of the system. In other words, at what point does a loss in function constitute a hazard to the animal? Thus, even the most carefully considered approach to the study has shortcomings, and findings are not easily generalized.

In the only study on bowhead whale baleen, Braithwaite (1983) used horizontally mounted plates to filter brine shrimp, *Artemia salina*, from a volume of chilled water discharged onto the upper fringed margins of the plates. Flow rates were established as the maximum volume that could pass through the baleen filter preparation in a 1-minute test. Water pressure was curiously low; the system was gravity-fed by a constant water column of only 7.5 cm over the baleen plates. After control values were established, the baleen hairs were brushed to uniform orientation, then light or medium crudes were brushed on to a nominal thickness of 0.5 to 1 mm, or poured to achieve a 1-cm-thick layer. An experimental run was considered valid only if the thickness of the oil coating remained uniform during the test. That was required for the purpose of analyzing data and was not intended to be a realistic portrayal of a fouling pattern. After the plates were fouled, the volume of water flowing through the preparation was measured and compared with control values. Brine shrimp were then introduced and water flow was measured again.

Details of the experimental protocol are not entirely clear. It appears that most of the 45 or more oil-fouling tests were performed with a single sample of baleen, with no information on whether control values were re-established following each successive fouling test. Results showed a 5 to 10% decrease in filtration rate after the plates had been fouled.

We used another approach to evaluate the fouling effects of oil (Geraci and St. Aubin, 1982, 1985). In a preliminary study, specimens from fin and gray whales were mounted in their natural hanging, or inverted position, in a continuous-flow water flume (Geraci and St. Aubin, 1982). Each preparation was oriented so that water flowed from the medial (lingual) to lateral (labial) surface, simulating the water expulsion–food retention phase of feeding. The flume contained freshwater at 15–20°C, and provided uniform flow with a velocity of 5–15 cm/second. Water movement was measured simultaneously along the incurrent and excurrent surfaces of the preparation and between the plates, using thermistor flowmeters. Once the flow pattern through a preparation was determined, the system was fouled either with a light or medium crude oil, or Bunker C. These were added to the water, and thus they struck the baleen preparation as a churned mixture, possibly in natural fashion. For some tests, roofing tar was applied directly to the fringed surface of the plates. Flow rates were again measured. Light to medium oils caused transient changes in water flow, which returned to normal within 40 seconds. Repeated oiling of the same preparation did not produce an additive effect. Bunker C had a more pronounced impact, restricting water flow for up to 15 minutes. Thereafter, though the plates were still noticeably fouled, normal flow patterns were recovered.

The study set the stage for a more detailed evaluation, using a system which allowed for testing in saltwater over temperatures ranging from 0 to 20°C and velocities up to 350 cm/second (Fig. 6–5). Samples from humpback and sei whales were tested along with new specimens from fin and gray whales. Pressure transducers were used to monitor water velocity at various points within the elliptical flume; resistance to flow could thus be calculated and used as the index of functional change. As expected from the pilot study, Bunker C had the greatest impact on water flow through baleen, particularly at temperatures of 0–5°C. Resistance to flow increased more than 100% in some humpback samples, and less than 75% in fin and sei whale preparations. Gray whale samples were relatively unaffected. Medium weight oil had little effect at any temperature.

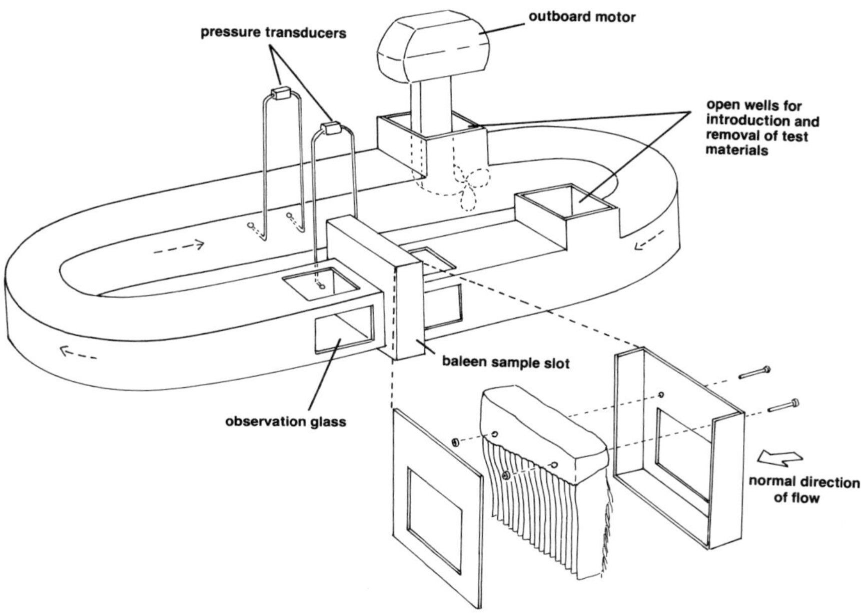

Figure 6-5
Tests were conducted in an elliptical tunnel with an outboard motor to circulate the water. Pressure transducers were positioned upstream and downstream from the baleen sample. Baleen specimens were mounted in a wooden frame for testing.

Selected samples were rinsed in continuously flowing saltwater for 32 hours; samples of baleen fibers were removed periodically and analyzed for residual oil (Fig. 6–6). Over 70% of the oil was lost within 30 minutes. In 8 of 11 trials, over 95% of the oil was cleared after 24 hours. We could not detect any change in resistance to flow in baleen after that time.

Combined evidence from the studies suggests that a spill of heavy oil, or residual patches of weathered oil, could interfere with the feeding efficiency of the fouled plates for several days at least. Effects would likely be cumulative in an animal feeding in a region so blanketed by weathered oil that the rate of cleansing is outpaced by fouling. That condition could describe the heart of a spill, or a contaminated bay or lead.

One can only speculate on consequences for a whale that occasionally eats a tar ball or engulfs a mouthful of weathered oil. The degree of fouling or damage required to impair feeding cannot be calculated with any precision, but in general, organs have some functional reserve. It seems that baleen does also; robust whales have been observed with damaged (Pivorunas, 1976) or rudimentary (Rice, 1961) plates. Judging from the low

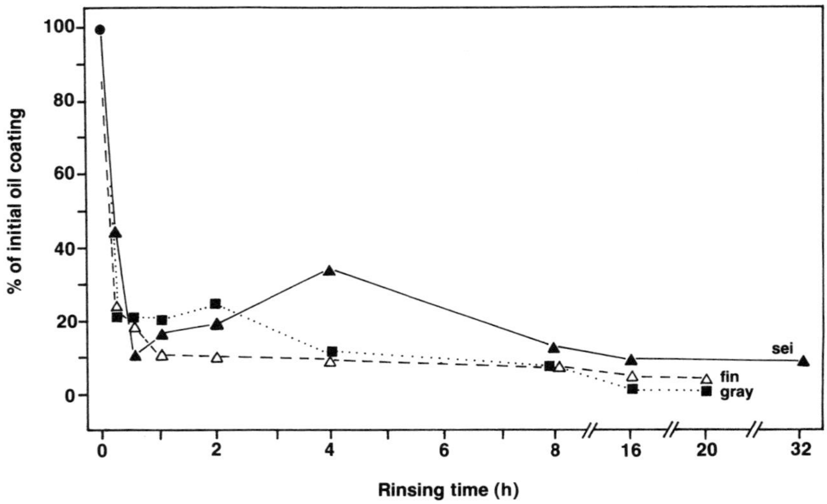

Figure 6-6
Amount of oil, as a percentage of initial coating, on fouled baleen preparations rinsed for up to 32 hours. Data are shown for one sample each from fin (△), sei (▲) and gray (■) whales. Over 70% of the oil was lost within 30 minutes. In 8 of 11 trials, over 95% was cleared after 24 hours.

level immediate impact in Braithwaite's (1981) study and the rate of clearance of oil in ours (Geraci and St. Aubin, 1985), it would appear that the concern for oiled whales is becoming less defensible. The cost and technology required to continue laboratory studies on fouling may not be worth the price.

Ingestion and Accumulation

Ingested petroleum hydrocarbons are toxic to many mammals. The subject evokes considerable interest because of the prevalence of, and devastating effects of, accidental hydrocarbon ingestion by young children (U.S. Department of Health and Welfare, 1978a,b). Oil compounds are systemically harmful, the degree depending on their chemical composition. Those with low viscosity and surface tension irritate the gastrointestinal tract and induce vomiting which leads to aspiration of the material into the lungs, causing pneumonia and death (Zieserl, 1979). Larger quantities, as much as 140 times the aspirated dose (Gerarde, 1964), can be tolerated if the substance remains in the gastrointestinal tract, but that is harmful as well.

Hydrocarbons can be directly toxic to the mucosal epithelium (Rowe *et al.*, 1973) and, when absorbed, travel throughout the body and produce their greatest effects on the central nervous system.

There has been some speculation that cetaceans could consume oil while feeding. Fraker *et al.* (1978) suggested that bowheads, because of their feeding behavior, could ingest damaging quantities of oil. Hansen (1985) affirmed that baleen whales that skim the surface and water column are more likely to ingest oil than gulp-feeders or toothed whales. Gray whales, because of their versatile feeding habits, could conceivably consume floating tar balls (Calkins, 1979) or contaminated bottom sediments (Hansen, 1985). Virtually any species might ingest oil by feeding on contaminated prey. The assumptions are logical, and one could fairly believe that of the vast quantities of oil discharged at sea (Neff, this volume, Chapter 1), at least a gulp or two must find its way into the gullet of a hungry whale. Yet these animals bear no evidence obvious enough to have drawn attention. As far as we know, the literature consists of only a threadbare notation that "hydrocarbons" were found in the intestines of two bottlenose dolphins along the coast of France (Duguy and Toussaint, 1977).

There has also been a study to determine how small quantities of refined petroleum oil consumed over a fairly long period of time would affect the health of *Tursiops* (Caldwell and Caldwell, 1982). It was an attempt to establish whether machine oil accidentally seeping into an aquarium pool might have been responsible for an unprecedented increase in mortality of captive dolphins. The only notable clinical finding had been elevated circulating levels of the enzyme glutamic pyruvic transaminase, suggesting that the liver might have been injured.

In the experiment, one dolphin was given the same machine oil in capsule form, at a rate of 5 mL/day, 5 days/week, for a total of 335 mL in 14 weeks. Another dolphin, used as a control, was given mineral oil under the same conditions, for a total of 225 mL in 10 weeks. The animals were examined for clinical, hematologic, and blood chemical changes during the study. None was found, nor was there any found during necropsy examination of the test dolphin when it was euthanized one month later for reasons unrelated to the study.

The results are not surprising. The quantity of machine oil consumed by the dolphin was substantially lower than the toxic dose for other mammals. Seals had also shown no effect after ingesting similar quantities (Geraci and Smith, 1976). In fact, the amount of substance considered to be critical is higher than one would reasonably wish to administer to a cetacean. In mice, it is in the order of 5 to 25 mL/kg for heavy fuel oils, and 14 to 20 mL/kg for lighter fuel oils (Elars Bioresearch Laboratories, Inc.,

1979a,b, 1980a–d). Let us assume that a cetacean would be at risk after taking a quantity of fuel oil at a midrange concentration of 15 mL/kg. To achieve that, an adult harbor porpoise would have to consume 1 L, a bottlenose dolphin 3–4 L, and a pilot whale 30 L. A forty-ton whale would require an estimated 600 L, or roughly 150 U.S. gal.

Could a cetacean in the wild ingest such quantities of oil? It is not feasible to predict the behavior of an excited animal unavoidably confronting a spill. It may swallow oil accidentally or, as we observe in terminally stressed odontocetes, drink seawater liberally, and with that, consume oil. Otherwise, it would seem unlikely, in the normal course of events, that a whale or dolphin would ingest much floating oil. A dolphin may drink 500 to 1500 mL of seawater daily (Ridgway, 1972). If contaminated, only a small portion of that would be oil. Odontocetes are predators that normally would not scavenge oil-killed fish, except perhaps for some bottlenose dolphins that have learned to forage behind fishing boats for a net-spilled meal. Lessons from captivity suggest that they would probably disregard tainted fish. Mysticetes in the area of a spill are more likely to ingest oil-contaminated food, particularly zooplankters which actively consume oil particles. Assuming toxic oils comprise 10% of the estimated 1600 kg of food consumed in a day by a 40-ton fin whale, the total quantity of ingested oil would be 160 kg. This approaches the critical dose calculated for highly toxic fuel oils. The question is, would fin whales feed around a spill of fresh volatile oil long enough to accumulate such quantities? There is no evidence from observational studies or stranding records to suggest that they do.

Petroleum hydrocarbons persist in the food chain, particularly in species that have a low capacity to detoxify them. Molluscs and other benthic invertebrates can accumulate residues from bottom sediments and remain contaminated for many years (Gilfillan and Vandermeulen, 1978). Gray whales and other bottom-feeding cetaceans might therefore ingest petroleum long after a spill has dissipated.

To predict the consequences of chronic ingestion of sub-lethal quantities of oil, we should know whether a cetacean can detoxify petroleum compounds or petroleum metabolites that persist in tissues of fish and other prey (McCain *et al.,* 1978). Cytochrome P-450, an iron-containing protein in liver cells, is part of a dynamic enzyme system involved in that metabolic process. It has been identified in liver from the bottlenose dolphin, white-sided dolphin, harbor porpoise, and minke whale (Geraci and St. Aubin, 1982; Goksoyr *et al.,* 1986), and is probably common to cetaceans generally. A pilot study on rats has shown that oil is a potent inducer of cytochrome P-450 (Geraci and St. Aubin, 1982), and we would expect it to have a comparable effect in a cetacean. These findings call for expanded studies on detoxifying systems in cetaceans. Animals suspected of having been exposed to oil should be analyzed for cytochrome P-450.

It is also possible to examine tissues for metabolites of petroleum hydrocarbons. We undertook a search for naphthalene in samples of liver and blubber from 15 species of whales which either had stranded, been taken in a fishery, or died in captivity. The analytical procedure was not particularly sensitive owing to limitations on methods for extracting naphthalene from tissue homogenates. Nevertheless, certain trends were evident. Highest levels were found in the blubber of small odontocetes; values in mysticetes were considerably lower. The pattern of accumulation seems to be consistent with the habitat of the animals. Beluga whales and narwhals, which had the highest concentrations in that study, live in a cold environment which retards hydrocarbon metabolism in fish (Collier *et al.*, 1978), potentially leaving more available to be consumed. Mysticetes generally feed on organisms that accumulate and eliminate hydrocarbons relatively rapidly (Neff *et al.*, 1976). Alternatively, the difference in the levels of naphthalene residues in odontocetes and mysticetes could reflect specific hydrocarbon detoxification capabilities in the two groups. These possibilities should be tested, and to do so, it will be necessary (1) to develop a sensitive method for analyzing naphthalene and its metabolites in marine mammal tissues; (2) to correlate levels with controlled ingestion of petroleum compounds; and (3) to analyze tissues from animals available through strandings or other opportunities. Data from these studies will provide the means to test the hypothesis that a cetacean may be harmed by ingested oil.

Summary

An oil spill at sea adds an element of risk to the environment of a whale or dolphin. Fresh crude oil or volatile distillates release toxic vapors that can damage sensitive tissues; harmful fractions may be swallowed or consumed through contaminated prey; and thicker tarry substances with entrapped debris may linger at the surface, plugging the vital baleen and digestive apparatus of whales that engulf them.

In spite of numerous observations of cetaceans in spills, none of these effects has been detected, or at least recorded with any certainty. Experimental evidence shows that dolphins can see oil at the surface and that they prefer to avoid it. Other cetaceans seem to be comparably equipped to detect oil. Yet in the wild, whales and dolphins have been observed swimming and feeding in its presence without apparent ill effect. Perhaps, in these instances, the stimulus was not noxious enough, or perhaps cetaceans disregard oil when they are engaged in more engrossing or important activities. Unlike furbearers, cetaceans do not lose heat through fouling of the skin. Furthermore, cetacean epidermis is nearly impenetrable, even to the highly volatile compounds in oil, and when skin is breached, exposure to these fractions does not impede the progress of healing. There is no evidence that oil or tar balls significantly foul the feeding apparatus of baleen whales; laboratory studies suggest that such fouling has only transient effects.

Current technology provides the means to probe deeper—to the molecular level, if necessary—for damage by oil to cetaceans. Probing may satisfy our scientific curiosity, but would not bring us closer to an understanding of the central question. On the whole, it is quite improbable that a species or population of cetaceans will be disabled by a spill at sea, whatever the likelihood that one or a few animals might be affected or even killed. Yet some habitats, and therefore their residents, are more vulnerable than others. The ice-edge, refuge for bowheads, narwhals, and beluga whales, is a riskier trap for them than pelagic waters. And in coastal areas with bustling oil production activity, dolphins might be the unwitting sentinels of a deteriorating environment. The stage is now set for decisions to identify, wisely utilize, and monitor such habitats.

References

Agarkov, G. B., Khomenko, B. G., and Khadzhinskiy, V. G. (1974). "Morphology of Delphinidae." Naukova-Dumka, Moscow (available JPRS 66036).

Albert, T. F. (1981). Some thoughts regarding the effect of oil contamination on bowhead whales, *Balaena mysticetus*. *In* "Tissue Structural Studies and other Investigations on the Biology of Endangered Whales in the Beaufort Sea." (T. F. Albert, ed.), Rep., Appendix No. 9, pp. 945–953. U.S. Bur. Land Manage., Washington, D.C.

Anonymous (1970). Santa Barbara oil spill. *Smithson. Inst., Cent. Short-Lived Phenom., Annu. Rep., 1969* Event No. 9-69, pp. 88–97.

Anonymous (1971). Alaska Peninsula oil spill. *Smithson. Inst., Cent. Short-Lived Phenom., Annu. Rep., 1970* Event No. 36-70. pp. 154–157.

Anonymous (1981). Oil spills off N.C. *Boston Globe*, May 9.

Anonymous (1982). Stranded, tarred pilot whale. *Smithson. Inst., Sci. Event Alert Network (SEAN) Bull.* **7**(3), 26.

Bass, M. (1986). Sniffing gasoline. *JAMA, J. Am. Med. Assoc.* **255**(19), 2604–2605.

Batelle Memorial Institute (1969). "Review of Santa Barbara Channel Oil Pollution Incident," Res. Rep., No. 14-12-530. Contract No. 14-12-530. Dept. of the Interior, Fed. Water Pollut. Control Admin. and Dept. of Transp., U.S. Coast Guard.

Bergey, M. (1979). "The *Ixtoc* Oil Spill—Effects on Marine Mammals and Turtles." Report by author aboard Res. Vessel Oregon II on 22-day shrimp tagging expedition.

Braithwaite, L. F. (1981). "The Effects of Oil on the Feeding Mechanism of the Bowhead Whale," Final Rep., Contract No. AA 851CT055. Brigham Young University, Provo, Utah.

Braithwaite, L. F. (1983). "The Effects of Oil on the Feeding Mechanism of the Bowhead Whale." Final Rep., Contract No. AA 851-CTO-55. U.S. Dept. of the Interior, Washington, D. C.

Brownell, R. L., Jr. (1971). Whales, dolphins and oil pollution. *In* "Biological and Oceanographic Survey of the Santa Barbara Channel Oil Spill 1969-1970" (D. Straughan, ed.), Sea Grant Publ. No. 2, Vol. 1, pp. 255–276. Allan Hancock Found. Univ. South. Calif. Los Angeles Ca.

Butler, M. J. A., Berkes, F., and Pawles, H. (1974). "Biological Aspects of Oil Pollution in the Marine Environment: A Review," Manuscr. Rep. No. 22. Mar. Sci. Cent., McGill University, Montreal.

Caldwell, M. C., and Caldwell, D. K. (1982). A study of the effects of oil ingestion on a bottlenose dolphin, *Tursiops truncatus*. *In* "Study of the Effects of Oil on Cetaceans" (J. R. Geraci and D. J. St. Aubin, eds.), Final Rep., Contract No. AA 551–CT9-29. pp. 224–237. U.S. Dept. of the Interior, Bur. of Land Manage., Washington, D. C.

Calkins, D. G. (1979). "Marine Mammals of Lower Cook Inlet and the Potential for Impact from Outer Continental Shelf Oil and Gas Exploration, Development and Transport," Final Rep., Outer Continental Shelf Environ. Assess. Program, Res. Unit 243, pp. 171-264. U.S. Dept. of Commerce, U.S. Dept. of the Interior, Anchorage, Alaska

Carpenter, C. P., Geary, D. L., Jr., Myers, R. C., Nachreiner, D. J.,Sullivan, L. J., and King, J. M. (1977). Petroleum hydrocarbon toxicity studies. XIV. Animal and human response to vapors of high aromatic solvent. *Toxicol. Appl. Pharmacol.* **41**(2), 235–249.

Collier, T. K., Thomas, L. C., and Malins, D. C. (1978). Influence of environmental temperature on disposition of dietary naphthalene in coho salmon (*Oncorhynchus kisutch*): Isolation and identification of individual metabolites. *Comp. Biochem. Physiol. C* **61C**, 23–28.

Cornish, H. H. (1980). Solvents and vapors. *In* "Casarett and Doull's Toxicology: the Basic Science of Poisons" (J. Doull, C. D. Klaassen, and M. O. Amdur, eds.)., 2nd ed., pp. 468–496. Macmillan New York.

Cowles, C. J., Hansen, D. J., and Hubbard, J. D. (1981). "Types of Potential Effects of Offshore Oil and Gas Development on Marine Mammals and Endangered Species of the Northern Bering Sea and Arctic Ocean," Tech. Pap No. 9. Bur. Land Manage., Alaska Outer Continental Shelf Reg., Anchorage.

Cummings, W. C., and Thompson, P. O. (1971). Gray whales, *Eschrichtius robustus*, avoid the underwater sounds of killer whales, *Orcinus orca*. *Fish. Bull.* **69,** 525–530.

Dawson, W. W., (1980). The cetacean eye. *In* "Cetacean Behavior: Mechanisms and Functions" (L. M. Herman, ed), pp. 53–100. Wiley, New York.

Duguy, R., and Toussaint, P. (1977). "Recherches sur les facteurs de mortalité des cétacés sur les côtes de France," C.M. 1977/N : 12, Cons. Int. Explor. Mer, Comité de Mammifères Marins.

Dutton, W. F. (1934). Petroleum dermatitis. *Med. Rec.* **140,** 550–552.

Easton, R., (1972). "Black Tide: The Santa Barbara Oil Spill and Its Consequences." Delcorte Press, New York.

Easton, R. (1972). "Black Tide: The Santa Barbara Oil Spill and its Consequences." Delcorte Press, New York.

Eberhardt, R. L., and Evans, W. E. (1962). Sound activity of the California gray whale, *Eschrichtius glaucus*. *J. Aud. Eng. Soc.* **10,** 324–328.

Elars Bioresearch Laboratories, Inc. (1979a). "Acute Toxicity Tests of API178-3 No. 2 Home Heating Oil (10% Cat)" Proj. No. 1443, API Med. Res. Publ. 2732773. Am. Pet. Inst., Washington, D. C.

Elars Bioresearch Laboratories, Inc. (1979b). "Acute Toxicity Tests of AP178-4 No. 2 Home Heating Oil (50% Cat)," Proj. No. 1443, API Med. Res. Publ. 2732772. Am. Pet. Inst., Washington, D. C.

Elars Bioresearch Laboratories, Inc. (1980a). "Acute Toxicity Tests of API 78-2 No. 2 Home Heating Oil (30% Cat)," Proj. No. 1443, API Med. Res. Publ. 2732771. Am. Pet. Inst., Washington, D. C.

Elars Bioresearch Laboratories, Inc. (1980b). "Acute Toxicity Tests of API 78-7 No. 6 Heavy Fuel Oil (API Gravity 17.1/0.8% S)," Proj. 1443, API Med. Res. Publ. 2732774. Am. Pet. Inst., Washington, D. C.

Elars Bioresearch Laboratories, Inc. (1980c). "Acute Toxicity Tests of API 79-2 No. 6 Heavy Fuel Oil (API Gravity 5.2/1.2% S)," Proj. 1443, API Med. Res. Publ. 2732813. Am. Pet. Inst., Washington, D. C.

Elars Bioresearch Laboratories, Inc. (1980d). "Acute Toxicity Tests of API 78-6 No. 6 Heavy Fuel Oil (API Gravity 11.7/2.7% S)," Proj. No. 1443, API Med. Res. Publ. 2732814. Am. Pet. Inst., Washington, D. C.

Evans, W. E. (1982). A study to determine if gray whales detect oil. *In* "Study of the Effects of Oil on Cetaceans" (J.R. Geraci and D.J. St. Aubin, eds.), Final Rep., Contract No. AA 551-CT9-29, pp. 47–61. U.S. Dept. of the Interior, Bur. Land Manage., Washington, D. C.

Fraker, M. A., Sergeant, D. E., and Hoek, W. (1978). "Bowhead and White Whales in the Southern Beaufort Sea," Beaufort Sea Proj. Techn. Rep. No. 4. Dept. of Fisheries and the Environment. Sidney, B.C.

Fritts, T.H., Irvine, A. B., Jennings, R. D., Collum, L. A., Hoffman, W., and McGehee, M. A. (1983). "Turtles, Birds, and Mammals in the Northern Gulf of Mexico and Nearby Atlantic Waters." Rep., Publ. FWS/OBS-82/65. U.S. Fish Wildl. Serv., Washington, D.C.

Geraci, J. R., and St. Aubin, D. J. (1980). Offshore petroleum resource development and marine mammals: A review and research recommendations. *Mar. Fish. Rev.* **42:** 1–12.

Geraci, J. R., and St. Aubin, D. J. (1982). "Study of the Effects of Oil on Cetaceans," Final Rep. U.S. Dept. of the Interior, Bur. Land Manage., Washington, D. C.

Geraci, J.R., and St. Aubin, D. J. (1985). "Expanded Studies of the Effects of Oil on Cetaceans," Final Rep., Part I. U.S. Dept. of the Interior, Minerals Manage. Serv., Washington, D. C.

Geraci, J. R., and Smith, T. G. (1976). Direct and indirect effects of oil on ringed seals (*Phoca hispida*) of the Beaufort Sea. *J. Fish. Res. Board Can.* **33**(9), 1976–1984.

Geraci, J. R., St. Aubin, D. J., and Reisman, R. J. (1983). Bottlenose dolphins, *Tursiops truncatus*, can detect oil. *Can. J. Fish. Aquat. Sci.* **40**(9), 1515–1522.

Gerarde, H. W. (1964). Kerosine—experimental and clinical toxicology. *Occup. Health Rev.* **16,** 17–21.

Giacometti, L. (1967). The skin of the whale (*Balaenoptera physalus*). *Anat. Rec.* **159,** 69–76.

Gilfillan, E. S., and Vandermeulen, J. H. (1978). Alterations in growth and physiology of soft-shell clams, *Mya arenaria*, chronically oiled with Bunker C from Chedabucto Bay, Nova Scotia, 1970-76. *J. Fish. Res. Board Can.* **35,** 630–636.

Goksoyr, A., Solbakken, J. E., Tarlebo, J., and Klungsoyr, J. (1986). Initial characterization of the hepatic microsomal P-450-system of the piked whale (minke) *Balaenoptera acutorostrata*. *Mar. Environ. Res.* **19,** 185–203.

Goodale, D. R., Hyman, M. A., and Winn, H. E. (1981). "Cetacean Responses in Association with the *Regal Sword* Oil Spill," Cetacean and Turtle Assess. Program, University of Rhode Island, Annu. Rep. 1979, pp. XI-1–XI-6. U.S. Dept. of the Interior, Washington, D. C.

Grose, P. L., and Mattson, J. S., eds. (1977). "The Argo Merchant Spill: A Preliminary Scientific Report." NOAA Environ. Res. Lab., Boulder, Colorado.

Gruber, J. A., (1981). Ecology of the Atlantic bottle-nosed dolphin (*Tursiops truncatus*) in the Pass Cavallo area of Matagorda Bay, Texas. M.Sc. Thesis, Texas A&M Univ., College Station.

Haldiman, J. T., Henk, W. G., Henry, R. W., Albert, R. W., Abdelbaki, Y. Z., and Duffield, D. W. (1985). Epidermal and papillary dermal characteristics of the bowhead whale (*Balaena mysticetus*). *Anat. Rec.* **211,** 391–402.

Hansbrough, J. F., Zapata-Sirvent, R., Dominic, W., Sullivan, J., Boswick, J., and Wang, X.-W. (1985). Hydrocarbon contact injuries. *J. Trauma* **25**(3), 250–252.

Hansen, D. J. (1985). "The Potential Effects of Oil Spills and Other Chemical Pollutants on Marine Mammals Occurring in Alaskan Waters," Rep. MMS 85-0031. U.S. Dept. of the Interior, Minerals Manage. Serv., Alaska Outer Continental Shelf Reg., Anchorage.

Herman, L. M., and Tavolga, W. N. (1980). The communication system of cetaceans. *In* "Cetacean Behavior: Mechanisms and Functions" (L. M. Herman, ed.), pp. 149–209. Wiley, New York.

Hunter, G. A. (1968). Chemical burns of the skin after contact with petroleum. *Br. J. Plast. Surg.* **21,** 337–341.

Ljungblad, D. K., Moore, S. E., and Van Schoik, D. R. (1983). "Aerial Surveys of Endangered Whales in the Beaufort, Eastern Chukchi, and Northern Bering Seas, 1982," Rep. No. NOSC TR1046. Alaska Outer Continental Shelf Reg., Minerals Manage. Serv., Anchorage.

Madsen, C. J., and Herman, L. M. (1980). Social and ecological correlates of cetacean vision and visual appearance. *In* "Cetacean Behavior: Mechanisms and Functions" (L. M. Herman, ed.), pp. 101–148. Wiley, New York.

Mann, G. (1946). Oyo y vision de las ballenas. *Biologica* **4,** 23–71.

McCain, B. B., Hodgins, H. O., Gronlund, W. D., Hawkes, J. W., Brown, D. W., Myers, M. S., and Vandermeulen, J. H. (1978). Bioavailability of crude oils from experimentally oiled sediments to English sole, *Parophrys vetulus* and pathological consequences. *J. Fish. Res. Board Can.* **35,** 657–664.

Neff, J.M., Cox, B. A., Dixit, D., and Anderson, J. W. (1976). Accumulation and release of petroleum-derived aromatic hydrocarbons by four species of marine animals. *Mar. Biol.* **38,** 279–289.

Nicol, C. W. (1976). "The Mizushima Oil Spill—A Tragedy for Japan and a Lesson for Canada," Rep. No. EPS-8-EC-76-2. Environ. Can. Environ. Prot. Serv., Ottawa, Canada

Ogawa, T., and Shida, T. (1950). On the sensory tubercles of lips and of oral cavity in the sei and fin whale. *Sci. Rep. Whales Res. Inst.* **3,** 1–16.

Owen, R. E. (1984). "Oil Spill from the British Tanker *Alvenas*." Natl. Mar. Fish. Serv., Southeast Fish. Cent., Miami, Florida (unpublished report).

Palmer, E., and Weddell, G. (1964). The relationship between structure, innervation, and function of the skin of the bottlenose dolphin, *Tursiops truncatus*. *Proc. Zool. Soc. London* **143,** 553–568.

Pepper, R. L., and Simmons, J. V. (1973). In-air visual acuity of the bottlenose dolphin. *Exp. Neurol.* **41,** 271–276.

Pike, G. (1962). Migration and feeding of the gray whale (*Eschrichtius gibbosus*). *J. Fish. Res. Board Can.* **19,** 815–838.

Pivorunas, A. (1976). A mathematical consideration on the function of baleen plates and hair fringes. *Sci. Rep. Whales Res. Inst.* **28,** 37–55.

Pivorunas, A. (1979). The feeding mechanisms of baleen whales. *Am. Sci.* **67**(4), 432–440.

Prieur, D., and Hussenot, E. (1978). Marine mammals stranded during the *Amoco Cadiz* oil spill. *Extrait de Penn ar Bed* **11**(94), 361–364.

Purves, P. E., and Pilleri, G. E., eds. (1983). "Echolocation in Whales and Dolphins." Academic Press, London.

Rice, D. W. (1961). Sei whales with rudimentary baleen. *Nor.Hvalfangst-Tid.* **5,** 189–193.

Rice, D. W., and Wolman, A. A. (1971). The life history and ecology of the gray whale, *Eschrichtius robustus. Am. Soc. Mammal., Spec. Publ.* No. 3.

Ridgway, S. H. (1972). Homeostasis in the aquatic environment. *In* "Mammals of the Sea: Biology and Medicine" (S. H. Ridgway, ed.), pp. 590–747. Thomas, Springfield, Illinois.

Rowe, L. D., Dollahite, J. W., and Camp, B. J. (1973). Toxicity of two crude oils and of kerosine to cattle. *J. Am. Vet. Med. Assoc.* **162**(1), 61–66.

St. Aubin, D. J., Geraci, J. R., Smith, T. G., and Friesen, T. G. (1985). How do bottlenose dolphins, *Tursiops truncatus*, react to oil films under different light conditions? *Can. J. Fish. Aquat. Sci.* **42**(3), 430–436.

St. Aubin, D. J., Stinson, R. H., and Geraci, J. R. (1984). Aspects of the structure and composition of baleen, and the effects of exposure to petroleum hydrocarbons. *Can. J. Zool.* **62,** 193–198.

Schmidt, R. F. (1977). Somatovisceral sensibility. *In* "Fundamentals of Sensory Physiology" (R. F. Schmidt, ed.), pp. 81–175. Springer-Verlag, New York.

Sergeant, D. E. (1970). Sea mammals and man: A look at the future. Fish. Res. Board Can., Arct. Biol. Stn., Ste. Anne de Bellevue, P.Q. (unpublished manuscript).

Shane, S. H., and Schmidly, D. J. (1978). "The Population Biology of the Atlantic Bottlenose Dolphin, *Tursiops truncatus*, in the Aransas Pass Area of Texas," Rep. No. MMC-76/11. U.S. Mar. Mammal Comm., Washington, D.C.

Slijper, E. J. (1979). *In* "Whales" (R. J. Harrison, ed), 2nd ed. Hutchinson, London.

Smith, T. G., Geraci, J. R., and St. Aubin, D. J. (1983). The reaction of bottlenose dolphins, *Tursiops truncatus*, to a controlled oil spill. *Can. J. Fish. Aquat. Sci.* **40**(9), 1522–527.

Sorensen, P. W., Medved, R. J., Hyman, M. A. M., and Winn, H. E. (1984). Distribution and abundance of cetaceans in the vicinity of human activities along the continental shelf of the northwestern Atlantic. *Mar. Environ. Res.* **12,** 69–81.

Spong, P., and White, E. (1971). Visual acuity and discrimination learning in the dolphin (*Lagenorhynchus obliquidens*). *Exp. Neurol.* **31,** 431–436.

Straughan, D. (1972). Biological effects of oil pollution in the Santa Barbara Channel. *In* "Marine Pollution and Sea Life." (M. Ruivo, ed.), pp. 355-359. Fishing News Ltd., London.

Thomson, C. A., and Geraci, J. R. (1986). Cortisol, aldosterone and leucocytes in the stress response of bottlenose dolphins, *Tursiops truncatus. Can. J. Fish. Aquat. Sci.* **43**(5), 1010–1016.

U.S. Dept. of Health and Welfare (1978a). "Tabulations of 1976 Case Reports." Publ. Health Serv./FDA. Bull. Natl. Clearinghouse Poison Control Centers, Bethesda, Maryland.

U.S. Dept. of Health and Welfare (1978b). "Phencyclidine Overdose Management." Publ. Health Serv./FDA. Bull. Natl. Clearinghouse Poison Control Centers, Bethesda, Maryland.

Valpey, R., Sumi, S. M., Copass, M. K., and Goble, G. J. (1978). Acute and chronic progressive encephalopathy due to gasoline sniffing. *Neurology* **28,** 507–510.

Waller, G. N. H. (1984). The ocular anatomy of cetacea—an historical perspective. *Invest. Cetacea,* **16,** 136–148.

Walls, G. L. (1963). "The Vertebrate Eye and its Adaptive Radiation." Hafner, New York.

Walsh, W. A., Scarpa, F. J., Brown, R. S., Ashcraft, K. W., Green, V. A., Holder, T. M., and Amoury, R. A. (1974). Gasoline immersion burn. *N. Engl. J. Med.* **291,** 830.

Wang, C. C., and Irons, G. U. (1961). Acute gasoline intoxication. *Arch. Environ. Health* **2,** 714–716.

Wolfram, M. A., Wolejsza, N. F., and Laden, K. (1972). Biomechanical properties of delipidized stratum corneum. *J. Invest. Dermatol.* **59**(6), 421–426.

Würsig, B., Clark, C. W., Dorsey, E. M., Fraker, M. A., and Payne, R. S. (1982). Normal behavior of bowheads. *In* "Behavior, Disturbance Responses and Feeding of Bowhead Whales, *Balaena mysticetus* in the Beaufort Sea, 1980-1981," pp. 33–143. LGL Ecol. Res. Assoc., Inc., Bryan, Texas/ U.S. Bur. Land Manage., Washington, D. C.

Yablokov, A. V., Bel'kovich, V. M., and Borisov, V. I. (1974). "Whales and Dolphins [Kity I Del'finy (1972)]." Izd. Nauka, Moscow/ Joint Publ. Res. Serv., Arlington, Virginia.

Zieserl, E. (1979). Hydrocarbon ingestion and poisoning. *Compr. Ther.* **5/6,** 35–42.

7

Sea Otters and Oil: Ecologic Perspectives

Katherine Ralls
Department of Zoological Research
National Zoological Park
Smithsonian Institution
Washington D.C. 20008

Donald B. Siniff
Department of Ecology, Evolution, and Behavior
University of Minnesota
Minneapolis, Minnesota 55455

Introduction

Otters comprise the subfamily Lutrinae of the family Mustelidae, order Carnivora. Most are associated with fresh water; only two species are associated with the marine environment. The marine otter, or chungungo, *Lutra felina*, which ranges along the west coast of South America, is somewhat smaller than its more extensively studied counterpart in the North Pacific, the sea otter, *Enhydra lutris* (Mason and MacDonald, 1986). The ensuing discussion will focus on the latter species.

The sea otter exploited the marine environment more recently in geologic time than did other mammals, and is consequently less specialized for life in the ocean. Its dense fur, streamlined shape, and amphibious habits are shared with its more terrestrial relative, the river otter. However, it possesses many aquatic adaptations (Estes, 1980), such as hindflippers, loss of the clavicle (which allows great flexibility of the pectoral girdle), and large kidneys. As a predator of marine invertebrates in the nearshore community, the sea otter possesses a modified dentition having strong, flattened premolars and molars well suited to crushing hard-shelled prey (Kenyon, 1969; Reidman and Estes, 1988).

The sea otter is the largest lutrinid (Mason and MacDonald, 1986) and the smallest marine mammal. An adult male can weigh up to 45 kg; females are considerably smaller (Estes, 1980).

Sea Mammals and Oil: Confronting the Risks
Copyright © 1990 by Academic Press, Inc.
All rights of reproduction in any form reserved.

Distribution and Abundance

The sea otter originally inhabited a large area in the North Pacific, ranging from the northern islands of Japan northward along the eastern coast of Kamchatka, through the Commander, Aleutian, and Pribilof islands to the Alaskan peninsula, and south along North America as far as central Baja California (Kenyon, 1969; Estes, 1980) (Fig. 7–1). However, the species was heavily exploited for its fur and was reduced to scattered, remnant populations by the beginning of the twentieth century. Some of these populations subsequently became extinct, but others increased when sea otters won legal protection under a number of statutes, beginning with the Treaty for the Preservation and Protection of Fur Seals in 1911 (Kenyon, 1969; Estes, 1980).

Sea otters now reoccupy most of their historic range across the North Pacific, from the southern end of the Kuril Islands eastward to Prince William Sound in the northeastern Gulf of Alaska (Reidman and Estes, 1988). In the 1960s and 1970s, Alaskan sea otters were translocated to southeast Alaska, British Columbia, Washington, and Oregon (Jameson

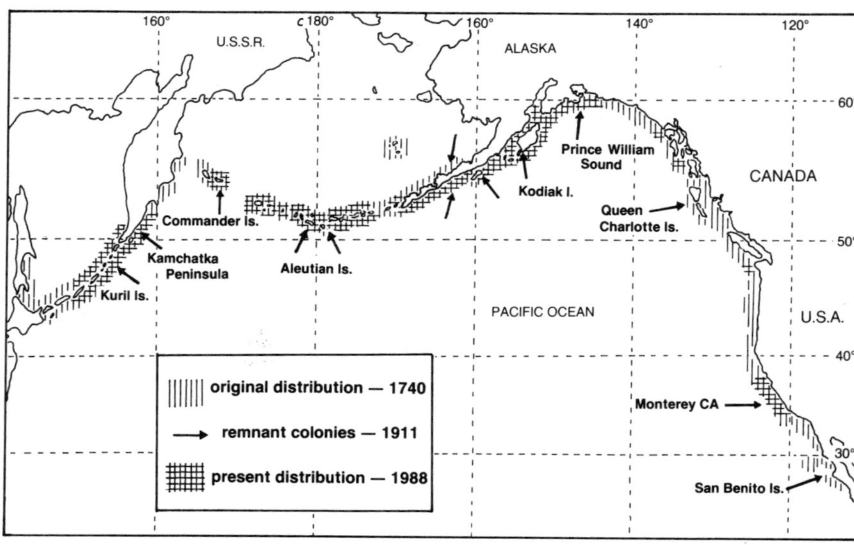

Figure 7-1
Distribution of the sea otter (Kenyon, 1969).

et al., 1982). The Oregon population died out, but the others became established and appear to be increasing. In the late 1970s, the total population was estimated to number at least 100,000 individuals, and perhaps as high as 150,000 (Kenyon, 1981).

The remnant population in California, which is classified as threatened by the U.S. Fish and Wildlife Service (Ladd, 1986a), also increased in size. It currently ranges along the coast between approximately Point Año Nuevo in the north and the Santa Maria River in the south, although some individuals wander beyond these limits (Leatherwood *et al.,* 1978). The central part of the California range is inhabited primarily by adult females and pups, along with juveniles and some territorial males; the most northern and southern areas are inhabited largely by males (Estes and Jameson, 1983). In 1987, the U.S. Fish and Wildlife Service translocated sea otters to another part of their historic range, San Nicolas Island off southern California (Ladd, 1986a,b; Brownell, 1988). The California population contains 1800 to 2000 individuals (Kenyon, 1981).

Preferred Habitat

Sea otters are primarily a coastal species. However, in areas where the water is shallow, large numbers of individuals sometimes occur far from shore. They are frequently found over 8 km from shore off the Copper River Delta in the Gulf of Alaska (Monnett, 1987) and over 30 km from shore around Unimak Island in the Bering Sea (Kenyon, 1969). Sea otters prefer to rest in kelp, as it helps keep them from drifting and perhaps affords some protection from white sharks (*Carcharodon carcharias*). In California, sea otters are most commonly observed inshore of the outer limit of the giant kelp (*Macrocystis pyrifera*) canopy (Reidman and Estes, 1988). However, juvenile males spend much of their time in open water beyond the limits of the kelp (Siniff and Ralls, 1988), and otters are abundant in many areas, such as Prince William Sound, Alaska, where there is very little kelp.

Sea otters occur over a variety of substrate types and nearshore communities. In California, areas with rocky bottoms support higher densities of otters (approximately 5 individuals/km^2) than those with sandy bottoms (0.8 individuals/km^2) (California Department of Fish and Game, 1976). This difference indicates that rocky bottoms are more productive communities and support a greater variety of invertebrate prey species taken by sea otters (Reidman and Estes, 1988).

Site Fidelity and Movement Patterns

Individual sea otters typically remain within a small area from day to day. In California, the average distance between successive daily locations, determined by radiotracking, ranged from 0.7 km for adult males to 2.4 km for juvenile males (Siniff and Ralls, 1988). Movements of over 10 km per day are infrequent, but in both California and Alaska, otters sometimes move over distances of 100 km or more within a few days. Thus, home ranges consist of heavily used areas connected by travel corridors. Individuals can also range over extended areas (Ribic, 1982a; Garshelis and Garshelis, 1984; Monnett and Rotterman, 1987; Siniff and Ralls, 1988). During a 2-year study period in California adult females visited an average of 24 km of coastline, juvenile females 47 km, adult males 98 km, and juvenile males 128 km (Siniff and Ralls, 1988).

In Alaska, sea otters often make long-distance seasonal movements between breeding and wintering areas (Garshelis and Garshelis, 1984; Monnett and Rotterman, 1987). In California, males may (Estes and Jameson, 1983; Jameson, 1989) or may not (Loughlin, 1980; Siniff and Ralls, 1988) show similar behavior; females in California do not move seasonally.

An otter's movements appear to be governed by its preference for specific kelp beds and feeding areas (Loughlin, 1977). Otters of all ages and both sexes, including territorial males, sometimes travel for long distances and then return to the same area (Loughlin, 1977; Estes and Jameson, 1983; Garshelis and Garshelis, 1984; Garshelis *et al.*, 1984; Siniff and Ralls, 1988). At least 5 of 17 male California sea otters that were tagged and moved a distance of 72 km returned to the vicinity of their capture location within 9 months (Odemar and Wilson, 1969; Wild and Ames, 1974). We observed two males in California that were moved approximately 150 km. One returned within a week and the other in about 1 month.

Reproduction

Sea otters can mate and pup throughout the year; reproductive activities are more seasonal in Alaska than in California. In Prince William Sound, mating generally occurs in fall and most pups are born in May (Garshelis *et al.*, 1984). The same pattern probably exists throughout Alaska, although the pupping peak may extend into summer in some areas, with mating activity common as late as December (Kenyon, 1969; Schneider, 1973). In the western Pacific, there may be two reproductive peaks, one in June–July and the other in September–October (Barabash-Nikiforov *et al.*, 1968). In California, there is a pupping peak in spring and a smaller, secondary one in fall (Siniff and Ralls, 1988).

In California, the period from copulation to birth is about 6 months (Estes and Jameson, 1983), and is thought to include a variable period of delayed implantation (Kenyon, 1981). Kenyon (1969) estimated the gestation period in Alaska to be about 120 days. Birth may occur on land (Jameson, 1983) but aquatic births—at least in California—are thought to be more common. Although twins are reported, they are apparently never successfully raised (Jameson and Bodkin, 1986). Most pups are weaned within 5–8 months (Garshelis *et al.*, 1984; Payne and Jameson, 1984; Wendell *et al.*, 1984; Monnett, 1987). After about 4 months of age, the pup's diet consists mostly of prey captured by its mother, although the pup continues to suckle until they separate (Payne and Jameson, 1984).

Female sea otters reach sexual maturity at 3–5 years of age in Alaska (Kenyon, 1969; Garshelis *et al.*, 1984). There are no comparable data for California, except for the single observation of a pup produced by a 4-year-old female (Wendell *et al.*, 1984). Males are thought to reach sexual maturity at 5 or 6 years of age (Green, 1978; Schneider, 1978) but generally begin breeding at a somewhat older age. Garshelis (1984) found that none of the breeding males sampled from Prince William Sound was younger than 6 years.

Early studies of reproductive tracts from animals collected in the Soviet Union and Alaska suggested that sea otters generally give birth only once every 2 years (Barabash-Nikiforov *et al.*, 1968; Kenyon, 1969; Schneider, 1973). However, subsequent observations of tagged individuals showed that most females pup annually (Jameson and Johnson, 1979; Loughlin *et al.*, 1981; Estes and Jameson, 1983; Garshelis *et al.*, 1984; Wendell *et al.*, 1984; Siniff and Ralls, 1988).

Some sea otter populations in Alaska have increased at rates as high as 16% per year (Kenyon, 1969; Estes 1982). Annual increment of the California population is currently above 5% (Jameson and Estes, 1988), a rate which characterized the population for many decades (Ralls *et al.*, 1983; Brody, 1987) except for a 10–15 year period beginning in the early 1970s when there was no apparent increase (Estes and Jameson, 1983; Wendell *et al.*, 1986; Brody, 1987). The population dynamics of sea otters are not well understood. A recently developed computer model may provide some insights (Brody, 1987).The relevance of modeling to predicting recovery rates after oil spills is addressed in Chapter 12 in this volume.

Social Organization and Reproductive Behavior

Sea otters are polygynous. Breeding males defend territories seasonally or year round in "female areas" occupied by reproductive females and their young (Kenyon, 1969; Calkins and Lent, 1975; Loughlin, 1980; Estes and Jameson, 1983; Garshelis *et al.,* 1984; Jameson, 1989). A male usually forms a temporary consortship, lasting 1–4 days, with an estrous female (Kenyon, 1969; Vandevere, 1970; Garshelis *et al.,* 1984). The two remain close together during this period. Males may attend females with large pups, or even some with small pups, although copulation is rarely successful until after the nursing period is over (Kenyon, 1969; Calkins and Lent, 1975; Garshelis *et al.,* 1984). Adult females do not always remain in the territory of a single male.

Large numbers of males congregate in "male areas" where few reproductive females are present. The groups include individuals too young to breed and older males when these are not maintaining seasonal territories. Male otters are usually the first to colonize new habitat, so male areas tend to be located at the edge of the range, where food is more abundant (Benech, 1981; Estes and Jameson, 1983; Garshelis *et al.,* 1984, 1986). Otters often rest in groups called "rafts," their composition fluid. Groups consisting primarily of females are usually small in California and Prince William Sound, although aggregations of more than 100 females have been observed in parts of Alaska (Monnett and Rotterman, 1987). Throughout this range, males may rest in very large rafts consisting of hundreds of individuals (Kenyon, 1969; Garshelis *et al.,* 1984). Otters sometimes emerge from the water to rest on ice or land; they do this more frequently, and in larger groups, in Alaska and the U.S.S.R. than in California (Kenyon, 1969; Faurot, 1985), perhaps because of the more severe climate and less human disturbance in northern waters.

Mothers and pups have strong social bonds, remaining together until the pup is weaned (Payne and Jameson, 1984); however, mothers that are ill or in poor condition may abandon pups (Garshelis and Garshelis, 1987). Weaning appears to be abrupt. In Prince William Sound, most male weanlings move long distances soon after they separate from their mothers. They establish new home ranges in male areas; females tend not to travel as far and remain within the natal female area (Monnett and Rotterman, 1987). Adult females sometimes reassociate with weaned pups (Garshelis and Garshelis, 1987); nothing is known about the frequency of this behavior or the extent to which female otters might associate with older related individuals.

Behaviors associated with reproduction can be injurious and even fatal (Garshelis *et al.*, 1984). While copulating, or attempting to, the male secures a bite-hold on the female's nose or face (Kenyon, 1969; Vandevere, 1970), sometimes leaving her with bloody lesions (Brosseau *et al.*, 1975) and lasting scars (Foott 1971). Although territorial males occasionally fight (Fisher, 1939; Vandevere, 1970; Loughlin, 1977), the frequency and extent of resulting injuries are unknown.

Trophic Levels and Feeding Habits

Feeding otters alternately dive to obtain prey and float at the surface to consume it. Pups less than about 6 weeks old are unable to dive and must remain on the surface while their mothers search for food (Payne and Jameson, 1984). The use of rocks as tools to dislodge or break open hard-shelled prey is more common in California (Hall and Schaller, 1964) than in Alaska (Kenyon, 1969).

Most sea otters feed almost entirely on macroinvertebrates, although in some areas they also consume epibenthic fish (Estes *et al.*, 1981; Maminov and Shitkov, 1970). At the population level, the diet becomes more varied after sea otters have occupied an area for several years (Estes *et al.*, 1981; Garshelis *et al.*, 1986; Reidman and Estes, 1988). Typically, after colonizing a new area, most of the otters consume large, calorically rich prey items. As these become scarcer, the diet begins to include smaller items and less preferred species. Individual otters show marked variation in food preference; one individual may feed largely on clams and another on mussels and turban snails (Estes *et al.*, 1981; Lyons and Estes, 1985; Siniff and Ralls, 1988). These differences persist over long periods of time (Lyons, 1987). Individuals also differ with respect to average dive length, time at the surface, feeding-bout length, and the interval between feeding bouts (Siniff and Ralls, 1988).

In the wild, otters can feed as much as 12 hours each day (Garshelis *et al.*, 1986; Siniff and Ralls, 1988). In California, juvenile females seem to spend more time feeding than males or females of other age classes, and often feed during the day, when other otters are resting (Siniff and Ralls, 1988). As a group, otters show no particular preference for day or night feeding, though they are most active during late afternoon, early evening, and early morning. Individuals differ in their diurnal feeding patterns, perhaps in association with the accessibility of their prey. The interval between feeding bouts is about 3 hours, which is the time it takes for food to be digested and cleared from the gut (Stulken and Kirkpatrick, 1955; Costa, 1978; Ribic, 1982b; Siniff and Ralls, 1988).

Metabolism and Thermoregulation

The otter's rapt attention to feeding directly reflects its need to fuel a metabolic rate which is over twice that of a terrestrial mammal of comparable size (Morrison *et al.,* 1974; Costa and Kooyman, 1982; Davis *et al.,* 1988). Captive animals consume 190–250 kcal/kg/day (Estes, 1980). There are no comparable values for wild otters, but Costa (1978) estimated that they must consume the equivalent of 23–33% of their body weight per day. Sea otters are locked into a feeding mode which cannot be interrupted.

Unlike all other marine mammals, the sea otter has little subcutaneous fat to aid in reducing heat loss; it must depend on its dense pelage for insulation. The pelage consists of an outer layer of guard hairs and an extremely dense underfur with over 100,000 hairs per square cm (Kenyon, 1969). The hairs are arranged in bundles, each containing approximately one guard hair and 70 underfur hairs. Bundle density varies over the body, ranging from about 2400 per square cm on the forearm to 851 on the dorsal surface. Air is trapped within the fur, adding to its insulative value (Kenyon, 1969; Costa and Kooyman, 1982; Davis *et al.,* 1988).

The pelage must be continually groomed to be effective. Loughlin (1977) described a somewhat stereotyped sequence of grooming. After an initial period of vigorous rolling and somersaulting, the otter rubs various parts of its body with the forepaws. It then licks and rubs its tail, hindquarters, chest, and forepaws. Sea otters typically groom themselves rather than each other. However, young pups are unable to groom themselves, and a female may spend up to 20% of the daylight hours grooming her pup (Sandegren *et al.,* 1973).

Summarizing the Risk

Many features of the life history of otters predispose them to exposure to spilled oil. They are coastal animals with strong site fidelity; they feed on benthic organisms that accumulate and store hydrocarbon residues from contaminated sediments (Neff, this volume, Chapter 1); and they spend a great deal of time at the surface moored to kelp beds that capture oil. The risk is amplified by circumstances that bring otters together to form loose associations, or rafts, sometimes numbering a hundred or more individuals. Under those circumstances, large groups of animals can be exposed simultaneously.

Isolated populations, such as the one along the coast of California, are most endangered. As we have learned from the 1989 *Exxon Valdez* accident in Prince William Sound, a regional disaster can kill many animals. The prognosis is not necessarily bleak, however, because "seed" or "pioneer" populations from remote areas can spread and recolonize, as they did after the near-extinction of sea otters at the turn of the century. And we now know that it is sometimes feasible to translocate otters to unoccupied areas where they once thrived.

For the individual otter, contamination by oil spells an almost certain fate. An otter's normal preoccupation with grooming becomes obsessive, displacing other behaviors, including feeding, resting, and caring for young. Despite its efforts, the otter may not be able to restore the insulative value of its coat.

References

Barabash-Nikiforov, I. I., Marakov, S. V., and Nikolaev, A. M. (1968). "Otters." Izd. Nauka, Leningrad.

Benech, S. V. (1981). "Observations of the Sea Otter *Enhydra lutris* Population Between Point Buchon and Rattlesnake Creek, San Luis Obispo, California, January through December 1980," VIII-6-81 Ecomar, Inc., Santa Barbara, California.

Brody, A. (1987). A spatially explicit simulation model of the California sea otter population. *In* "Population Status of California Sea Otters" (D. B. Siniff and K. Ralls, eds.), Final Rep. Contract No. 14-12-001-3003. Minerals Manage. Serv., Los Angeles, California.

Brosseau, C., Johnson, M. L., Johnson, A. M., and Kenyon, K. W. (1975). Breeding the sea otter at Tacoma Aquarium. *Int. Zoo Yearb.* **15,** 144-147.

Brownell, R. L., Jr. (1988). California sea otter translocation. *Mar. Mammal. Sci.* **4,** 85.

California Department of Fish and Game (1976). "A Proposal for Sea Otter Protection and Research, and Request for Return of Management to the State of California," 2 vols. (unpublished report).

Calkins, D. G., and Lent, P. C. (1975). Territoriality and mating behavior in Prince William Sound sea otters. *J. Mammal.* **56,** 528–529.

Costa, D. P. (1978). The ecological energetics, water, and electrolyte balance of the California sea otter, *Enhydra lutris*. Ph.D. Thesis, University of California, Santa Cruz.

Costa, D. P., and Kooyman, G. L. (1982). Oxygen consumption, thermoregulation, and the effect of fur oiling and washing on the sea otter, *Enhydra lutris*. *Can. J. Zool.* **60,** 2761–2767.

Davis, R. W., Williams, T. M., Thomas, J. A., Kastelein, R. A., and Cornell, L. H. (1988). The effects of oil contamination and cleaning on sea otters (*Enhydra lutris*). II. Metabolism, thermoregulation, and behavior. *Can. J. Zool.* **66,** 2782–2790.

Estes, J. A. (1980). *Enhydra lutris*. *Mammal. Species* **133,** 1–8.

Estes, J. A. (1982). The case of the sea otter. *In* "Problems in Management of Locally Abundant Wild Mammals" (P. A. Jewell, S. Holt, and D. Hart, (eds.), pp. 167–180. Academic Press, New York.

Estes, J. A., and Jameson, R. J. (1983). "Summary of Available Population Information on California Sea Otters," POCS Tech. Pap. No. 83–11, Interagency Agreement No. 14-12-001 Minerals Manage. Serv. Washington, D.C.

Estes, J. A., Jameson, R. J., and Johnson, A. M. (1981). Food selections and some foraging tactics of sea otters. *In* "Proceedings of the Worldwide Furbearer Conference, 1980" (J. A. Chapman and D. Pursley, eds.), pp. 606–641. Frostburg, Maryland.

Faurot, E. (1985). Haulout behavior of California sea otters, *Enhydra lutris*. *Mar. Mammal. Sci.* **1,** 337–340.

Fisher, E. M. (1939). Habits of the southern sea otter. *J. Mammal.* **20,** 21–36.

Foott, J. O. (1971). Nose scars in female sea otters. *J. Mammal.* **51,** 621–622.

Garshelis, D. L. (1984). Age estimation of living sea otters. *J. Wildl. Manage.* **48**(2), 456–463.

Garshelis, D. L., and Garshelis, J. A. (1984). Movements and management of sea otters in Alaska. *J. Wildl. Manage.* **48,** 665–678.

Garshelis, D. L., and Garshelis, J. A. (1987). Atypical pup rearing strategies by sea otters. *Mar. Mammal. Sci.* **3**(3), 263–270.

Garshelis, D. L., Johnson, A. M., and Garshelis, J. A. (1984). Social organization of sea otters in Prince William Sound, Alaska. *Can. J. Zool.* **62,** 2648–2658.

Garshelis, D. L., Garshelis, J. A., and Kimker, A. T. (1986). Sea otter time budgets and prey relationships in Alaska. *J. Wildl. Manage.* **50,** 637–647.

Green, B. (1978). Sexual maturity and senescence of the male California sea otter (Enhydra lutris). M.Sc. Thesis, San Jose State University, San Jose, California.

Hall, E. R., and Schaller, G. B. (1964). Tool-using behavior of the California sea otter. *J. Mammal.* **45,** 287–298.

Jameson, R. J. (1983). Evidence of birth of sea otter on land in central California. *Calif. Fish Game* **69,** 122–123.

Jameson, R. J. (1989). Movements, home range, and territories of male sea otters off central California. *Mar. Mammal. Sci.* **5,** 159–172.

Jameson, R. J., and Estes, J. A. (1988). Status of the California sea otter population. Abstr., *Am. Soc. Mammal. Annual Meetings,* Clemson, South Carolina.

Jameson, R. J., and Bodkin, J. L. (1986). An incidence of twinning in the sea otter (*Enhydra lutris*). *Mar. Mammal. Sci.* **2,** 305–309.

Jameson, R. J., and Johnson, A. M. (1979). Evidence of annual reproduction among sea otters. *Abstr., 3rd Bienn. Conf. Biol. Mar. Mammal.,* Seattle, Washington.

Jameson, R. J., Kenyon, K. W., Johnson, A. M., and Wight, H. M. (1982). History and status of translocated sea otter populations in North America. *Wildl. Soc. Bull.* **10,** 100–107.

Kenyon, K. W. (1969). The sea otter in the eastern Pacific Ocean. *North Am. Fauna* **68,** 1–352.

Kenyon, K. W. (1981). Sea otter, *Enhydra lutris*. *In* "Handbook of Marine Mammals" (S. H. Ridgway and R. J. Harrison, eds.), vol. 1, pp. 209–223. Academic Press, New York.

Ladd, W. (1986a). New hope for the Southern sea otter. Part 1. *End. Spec. Tech. Bull.* **11**(8 & 9), 12–14.

Ladd, W. (1986b). New hope for the Southern sea otter. Part 2. *End. Spec. Tech. Bull.* **11**(10 & 11), 5–7.

Leatherwood, S., Harrington-Coulombe, L. J., and Hubbs, C. L. (1978). Relict survival of the sea otter in central California and evidence of its recent redispersal south of Point Conception. *Bull. South. Calif. Acad. Sci.* **77,** 109–115.

Loughlin, T. R. (1977). Activity patterns, habitat partitioning, and grooming behavior of the sea otter, *Enhydra lutris*, in California. Ph.D. Thesis, University of Califalifornia, Los Angeles.

Loughlin, T. R. (1980). Radio telemetric determination of the 24-hour feeding activities of sea otters, *Enhydra lutris*. *In* "A Handbook on Biotelemetry and Radio Tracking" (C. J. Amlaner, Jr. and D. W. MacDonald, eds.), pp. 717–724. Pergamon, Oxford.

Loughlin, T. R., Ames, J. A., and Vandevere, J. E. (1981). Annual reproduction, dependency period and apparent gestation period in two California U.S.A. sea otters *Enhydra lutris. Fish. Bull.* **79,** 347–349.

Lyons, K. J., (1987). Individual variation in the diet and foraging strategy in the female California sea otter, *Enhydra lutris. Abstr., Anim. Behav. Soc., 1987,* Williamstown, Massachusetts.

Lyons, K. J., and Estes, J. A. (1985). Individual variation in diet and foraging strategy in the female California sea otter, *Enhydra lutris. Abstr. West. Soc. Nat. 66th Annu. Meet.,* Monterey, California.

Maminov, M. K., and Shitkov, A. M. (1970). On the food of the sea otter on Paramushir Island. *Proc. Pac. Res. Inst. Fish Ocean. (TINRO)* **70,** 176–179.

Mason, C. F., and MacDonald, S. M. (1986). "Otters: Ecology and Conservation." Cambridge Univ. Press, London and New York.

Monnett, C. (1987). Movement, developmental and mortality patterns of sea otters in Alaska. Ph.D. Thesis, University of Minnesota, Minneapolis.

Monnett, C., and Rotterman, L. (1987.) Movement patterns of adult female and weanling sea otters in Prince William Sound, Alaska. *In* "Population Status of California Sea Otters" (D. B. Siniff and K. Ralls, eds.), Final Rep. Contract No. 14-12-001-3003, pp. 193–235. Minerals Manage. Serv., Los Angeles, California.

Morrison, P. M., Rosenmann, M., and Estes, J. A. (1974). Metabolism and thermoregulation in the sea otter. *Physiol. Zool.* **47,** 218–229.

Odemar, M. W., and Wilson, K. C. (1969). Results of sea otter capture, tagging and transporting operations by the California Department of Fish and Game. *Proc. 6th Annu. Conf. Biol. Sonar Diving Mammals,* Stanford Res. Inst., Menlo Park, California, pp. 73–79.

Payne, S. F., and Jameson, R. J. (1984). Early behavioral development of the sea otter, *Enhydra lutris. J. Mammal.* **65,** 527–531.

Ralls, K., Ballou, J., and Brownell, R. L., Jr. (1983). Genetic diversity in California sea otters: Theoretical considerations and management implications. *Biol. Conserv.* **25,** 209–232.

Reidman, M., and Estes, J. (1988). A review of the history, distribution and foraging ecology of sea otters. *In* "The Community Ecology of Sea Otters" (G. R. Van Blaricom and J. A. Estes, eds.), pp. 4–21. Springer-Verlag, Berlin.

Ribic, C. A. (1982a). Autumn movement and home range of sea otters in California. *J. Wildl. Manage.* **46,** 795–801.

Ribic, C. A. (1982b). Autumn activity of sea otters in California. *J. Mammal.* **63,** 702–706.

Sandegren, F. E., Chu, E. W., and Vandevere, J. E. (1973). Maternal behavior in the sea otter. *J. Mammal.* **54,** 668–679.

Schneider, K. (1973). "Reproduction in the Female Sea Otter," Prog. Rep. Federal Aid in Wildl. Restoration Proj. W-17-4 and W-17-5. Alaska Dept. of Fish and Game, Anchorage.

Schneider, K. (1978). "Reproduction in the Female Sea Otter in the Aleutian Islands" (unpublished rep.). Alaska Dept. of Fish and Game, Anchorage.

Siniff, D. B., and Ralls, K., eds. (1988). "Population Status of California Sea Otters," Final Rep. Contract No. 14-12-0013003. Minerals Manage. Serv., Los Angeles, California.

Stulken, D. E., and Kirkpatrick, C. M. (1955). Physiological investigation of captive mortality in the sea otter (*Enhydra lutris*). *Trans. North Am. Wildl. Conf.* **20,** 476–494.

Vandevere, J. E. (1970). Reproduction in the southern sea otter. *Proc. 7th Annu. Conf. Biol. Sonar Diving Mammals,* Stanford Res. Inst., Menlo Park, California, pp. 221–227.

Wendell, F. E., Ames, J. A., and Hardy, R. A. (1984). Pup dependency period and length of reproductive cycle: Estimates from observations of tagged sea otters, *Enhydra lutris,* in California. *Calif. Dep. Fish Game* **70,** 89–100.

Wendell, F. E., Hardy, R. A., and Ames, J. A. (1986). "An Assessment of the Accidental Take of Sea Otters, *Enhydra lutris,* in Gill and Trammel Nets," Mar. Resour. Tech. Rep. No. 54. Calif. Dept. of Fish and Game, Sacramento.

Wild, P. W., and Ames, J. A. (1974). "A Report on the Sea Otter, *Enhydra lutris* L., in California," Mar. Resour. Tech. Rep. No. 20. Calif. Dept. of Fish and Game, Sacramento.

8

Physiologic and Toxic Effects on Sea Otters

Joseph R. Geraci
Department of Pathology
Ontario Veterinary College
University of Guelph
Guelph, Ontario
N1G 2W1 Canada

Thomas D. Williams
Monterey Bay Aquarium
Monterey, California 93940

Historical Notes

Oil spills are headline news wherever they occur, and increasingly there are attempts to search each incident for its biological effects. These can be so subtle as to escape detection or can unfold as a drama larger than the spill itself. Witness waterfowl, fatally drawn to a tranquil oil sheen, and sea otters, *Enhydra lutris,* driven to frenzied grooming by oil on their fur.

The otter is the marine mammal most likely to be harmed by oil, both immediately and in the long term. Kenyon (1969) offered the first relevant account when noting that the population of otters (estimated at 4000 in 1920) in the southern Shumagin Islands, Alaska, declined after a wrecked tanker and a freighter spilled oil there during World War II. Furthermore, Marakov (1963, cited in Kenyon, 1969) believed that human activities, including spills of petroleum products, prevented recovery of the otter population of the Commander Islands in the Bering Sea. Barbash-Nikiforov *et al.* (1968, cited in Van Blaricom and Jameson, 1982) gave an account of a spill of gasoline and diesel fuel in the nearby Kuril Islands, which spread along 40 km of the coastline and killed over 100 sea otters. The event occurred in 1946 (Howe, 1983), and for the next 20 years or so, there was only glancing mention of oil-fouled carcasses along the shores of Alaska (Anonymous, 1971; Hess and Trobaugh, 1971) and California (Kooyman *et al.*, 1976; Anonymous, 1980, 1982; Jameson,

Sea Mammals and Oil: Confronting the Risks
Copyright © 1990 by Academic Press, Inc.
All rights of reproduction in any form reserved.

1986). Then on March 24, 1989, the tanker *Exxon Valdez* grounded in Prince William Sound, releasing 11 million gal of Prudhoe Bay crude oil and ringing the death knell for nearly 1000 sea otters.

The first detailed account of the destructive potential of oil came from a spill affecting not the sea otter, but the related European otter, *Lutra lutra,* whose habitat includes nearshore waters along the western British Isles. Less than 2 months after becoming operational, the oil terminal at Sullom Voe in Shetland experienced its first significant mishap when on December 30, 1978, about 1200 tons of Bunker C fuel spilled from the tanker *Esso Bernicia* and eventually spread along 110 km of twisting shoreline (Richardson, 1979). After learning that river otters had been oiled, a team of investigators assessed the effect of the spill on the population and collected carcasses for necropsy examination (Baker *et al.,* 1981). One animal with a patchy coating of oil was taken for rehabilitation. It was cleaned, fed, and kept warm, and although it ate offerings of fish, it died two days later. Another was rescued but died the next day. By February 22, 1979, 11 other otters, partly or completely covered with oil, had been found dead. Necropsies were performed on five carcasses. Two had tracheitis; all had congested lungs; all had blood in the stomach; the intestines of four had blood mixed with oil. A limited examination of another otter revealed, in the intestines, bloody fluid which was shown by chromatographic analysis to contain oil.

The investigators believed that other river otters might have died unobserved in natural lairs and hiding places. Later casualties, not as heavily oiled, were discovered up to 150 m from the sea, where they had apparently sought shelter. Conclusion? At least 15%, and possibly as many as 50%, of the local river otters had been killed in the incident.

This event foreshadowed the blow dealt by the 1989 *Exxon Valdez* spill which fouled 1300–1600 km of coastline. The impact on sea otters was immediate and obvious. Many that were fouled came ashore and were easily captured, the first within 5 days of the spill. Those in the water were seized using dip nets, and were placed aboard fishing vessels where most of them died. There seemed to be no way to soften the impact; rescuers even suggested placing modified gill nets around the spill to prevent animals from reaching it. The plan was not implemented; instead, volunteers turned to rehabilitating sea otters in makeshift clinics.

At the time, the local population of otters was at least 4500 (Irons *et al.,* 1988). By mid-October, 886 carcasses had been recovered from the field; many more likely had died. This is the first chapter of the story, which may not be concluded until the sea otters repopulating the area no longer face the threat of contamination by residual oil.

Heavily matted with oil, the sea otter's pelt no longer provides life supporting insulation. Latouche Island, Prince William Sound, Alaska, 1989. Photo by Kathy Dawson.

Detection and Avoidance

The *Esso Bernicia* and *Exxon Valdez* events provide clear evidence of the fate of an otter that has encountered a substantial spill of oil. Earlier, Siniff and his colleagues (1982) had gained some understanding of the impact of oil from their study on two sea otters placed in a pool partially covered with Prudhoe Bay crude. Immediately after the oil was added, the otters became "nervous and curious," even though the oil was contained on the opposite side of a surface divider. Occasionally they would rear up and stare at the oil. They made determined efforts to escape. When swimming underwater around the perimeter of the pool, the otters would occasionally surface on the oiled side, but would remain for only a few seconds. They spent less than one minute of each hour on the oiled side, but eventually both became covered with oil. One would emerge from the water to the level of its shoulders and shake vigorously in an apparent attempt to rid itself of oil.

The study suggests that, given a choice, otters would avoid oil. Why then did the river otters not avoid the *Esso Bernicia* spill? Perhaps because of an overriding instinct to remain in the area. There is also some indication, from events surrounding the incident itself that, once fouled, an otter's preoccupation with grooming may make it inattentive to the threat of additional or repeated contact. Baker *et al.* (1981) had noted that, following the spill, the otters did not at first recognize oil as a danger. One otter dove and emerged in a patch of floating oil. Another walked along the rocky shore, making no apparent attempt to avoid regions with oil. Two more were seen on the shore; "one of these, rather atypically, did not appear to notice the observer, in full view, as he approached to within a meter. It was moderately oiled and intent on grooming . . . ."

The spill from the *Exxon Valdez* was so large as to provide no answers to the question of oil detection and avoidance. There was no uncontaminated water near the spill where the sea otters could find refuge. Otters were seen swimming through fingers of the slick, either unaware or unable to avoid them.

Yet otters should be able to detect oil, using vision perhaps (Limbaugh, 1961; Gentry and Peterson, 1967; Nachtigall, 1986), and smell (Elliot, 1875; Coues, 1877; Kenyon, 1969). As for smell, Barabash-Nikiforov *et al.* (1947) noted that otters react fairly vigorously to odors. They offered the example of Japanese poachers who fouled shore rocks with kerosene to frighten otters toward the sea, where they would be easier prey. It may be possible to test experimentally the ways by which they are able to sense oil. For practical purposes, it may be enough to project that otters are able to detect it. However, once contaminated, they are driven to compulsive grooming at a level of intensity that overpowers what might have been an avoidance response.

Surface Fouling

Grooming, an otter's key to survival, becomes a predictable weakness after the otter confronts oil. Siniff *et al.* (1982) brushed a coating of oil over a small area of the abdomen of four sea otters, then released them and used radio transmitters to follow their behavior in the wild. The animals spent most of the first few days grooming. Williams (1978), in an earlier version of the Siniff *et al.* (1982) report, had observed that after two otters contacted oil in a holding pool, they spent 75% of their time underwater trying to clean themselves. These observations were confirmed in subsequent studies (Costa and Kooyman, 1982; Davis *et al.*, 1988), leading Cimberg and Costa (1985) to conclude that otters are susceptible to slicks because they spend much time on the surface feeding, grooming, resting, and swimming.

Studies on muskrats (*Ondatra zibethicus*) (Wragg, 1954; McEwan *et al.*, 1974) foreshadowed the probable metabolic effects of surface fouling on sea otters. By destroying the insulative quality of the coat, oil disrupted the animals' sensitive energetic balance. Kooyman *et al.* (1977), after showing that the pelt of the sea otter was a superior insulator, demonstrated that oiling caused a twofold increase in thermal conductivity or loss of insulative value. They concluded that an animal unable to reverse such an effect by grooming could not endure water immersion.

Williams *et al.* (1988) approached the same question using more samples and a slightly different protocol. They found that pelts layered with fresh oil showed a threefold increase in thermal conductance. Weathered oil had less effect, presumably because it was more viscous and tended to remain on the tips of the guard hairs, outside the underfur.

Costa and Kooyman (1982) were the first to record the consequences of fouling to a live otter in a captive setting. They brushed low-viscosity crude oil over the back, covering about 18% of the animal's body surface. Three otters fouled in this way were then permitted to swim in a clean pool of water for 30 minutes. Their metabolic rates increased 40% over the control values obtained when they had been immersed in clean water. One animal, deliberately left oiled, displayed nearly a twofold increase in metabolic rate on the sixth day. Apparently, the otter had groomed, thereby spreading oil over the fur and further reducing its insulative value. Consumption of toxic oils likely compounded the metabolic stress, a pattern that also emerged in animals that were treated following the *Exxon Valdez* accident. When these damaging factors are combined, it becomes clear that, without help, an otter would not survive fouling of the entire body.

The same conclusion was drawn by Siniff *et al.* (1982) in their study of radio-tagged otters in Prince William Sound. The activity of the animals increased dramatically the first week after oiling. Most of it was associated with grooming; feeding patterns did not change. The otters survived the experiment because (the authors suggested) only a small portion of the pelage had been fouled, and food resources in the area were abundant.

A later study by Davis *et al.* (1988) identified, and for the first time graded under controlled conditions, each element of an otter's metabolic and behavioral responses to oil. The investigators measured metabolic rate before and after they fouled 20% of the fur with fresh crude oil. The result was a twofold increase in metabolic rate. This was slightly more than Costa and Kooyman (1982) had found and, by reason of better technique, was a more accurate reflection of the impact of fouling on energy balance. More important, however, was their observation that after fouling, grooming activity increased from 35% to 61%, swimming increased from 10% to 17%, and the time spent resting on the surface decreased from 49% to 12%. They also found that grooming activity, contrary to its "intent," made matters worse by spreading oil onto clean areas and rubbing it deep into the fur.

Grooming emerges as the common behavior driven by oil, whether in the natural setting following the *Esso Bernicia* or *Exxon Valdez* spills, or after deliberate fouling. It can be debated whether such an activity would be all consuming, as suggested by the studies of Williams (1978) and to a lesser extent by Davis *et al.* (1988), or whether animals under less confined, more natural conditions might continue their normal feeding patterns and ultimately recover from the experience (Siniff *et al.*, 1982).

It is a matter of degree. A more extensive coating of oil would likely have tipped the balance and delivered the otters in the Siniff *et al.* (1982) study into a tightening metabolic spiral: oil fouls the fur, reduces its insulative properties, and increases heat loss; the animal compensates by increasing its metabolic rate which, in turn, it must fuel by consuming more food; but eating gives way to vigorous grooming, and that energy, squandered on spreading the oil, is not restored; body mass decreases and more heat is lost.

One can assume that not all otters contacting oil will be affected to the same extent. Pups, by virtue of their size and the characteristically high density of their fur, would be most sensitive. So would animals foraging in marginal feeding areas, as they might have to expend greater energy to gather food needed to offset heat loss. Newly arriving otters that contact a fresh spill would be more vulnerable than latecomers, because the fractions that penetrate the pelt and dissolve natural lipids are found in fresh and not weathered oil. And otters with strong homing tendencies, initially frightened from a home site by a spill, might return to a less threatening scene. There are other complications, of course. Low sea temperatures increase heat loss and slow the weathering of oil; contaminated kelp beds prolong the time of contact; grooming of pups transfers oil from young to parents; and so on. In the broader sense, these all help to underscore the truth that the sea otter's vital commitment to grooming predetermines its fate once it has contacted oil.

Toxic Effects

A sea otter in or near fresh oil will inhale volatile vapors that injure the lungs and other organs and will ingest oil when it attempts to clean its fouled coat. After the grounding of the *Exxon Valdez,* intoxication was a major factor contributing to the death of sea otters. Those taken from the beaches showed signs of general depression (perhaps due to the direct effects of volatile hydrocarbons on the nervous system), respiratory distress associated with interstitial emphysema (rupture of air sacs), and diarrhea. Degenerative lesions in the liver were suggestive of a toxic condition; these could be diagnosed in the living animal by the finding of elevated levels of enzymes that leaked from the liver cells into the circulation. There was also evidence of kidney failure and endocrine imbalances affecting adrenal and thyroid glands. Ultimately, otters died of stress and shock.

More subtle changes might be expected as a consequence of ingesting prey items, like molluscs, which poorly metabolize petroleum hydrocarbons and will shuttle them unchanged to an otter (Neff, this volume, Chapter 1). We expect, without direct evidence, that otters have the same detoxifying enzyme system as pinnipeds, cetaceans, and mammals generally; the subject deserves further study. The net effect of ingesting hydrocarbons would, over the long term, impose additional stress to an animal or population that might have escaped or survived the initial impact of a spill.

Rehabilitating Fouled Otters

Attempts have been made to reverse the effects of fouling by cleaning an otter's pelt, in the same way birds have been treated after becoming coated with oil. From a series of studies, a protocol was eventually developed. Costa and Kooyman (1982) washed three oiled animals with 1–8% solutions of Amber Lux (Lever Brothers), an anionic surfactant-type detergent, then returned them to a holding pool at 20°C. The otters resumed their normal behavior and appearance, though one later died of pneumonia. One of the otters was oiled and washed again, and this time was placed in a holding area with water at 16°C. It shivered vigorously, but after being transferred to 25–30°C water, it groomed and recovered completely. Meanwhile, metabolic studies had shown that washing the otters had decreased the insulative value of the pelt by about 2–5 times. The effect was transitory; after about 8 days, the fur's thermal conductance had returned to normal.

Siniff *et al.* (1982) also washed an oiled otter, but with mixed results. After encountering crude oil in a small pool, an otter was cleaned with Polycomplex-11 (Guardian Chemical). The fur was wetted with cleaning agent, rinsed in water, and rubbed thoroughly with towels. The animal was then allowed to dry and groom for 8 hours before being placed

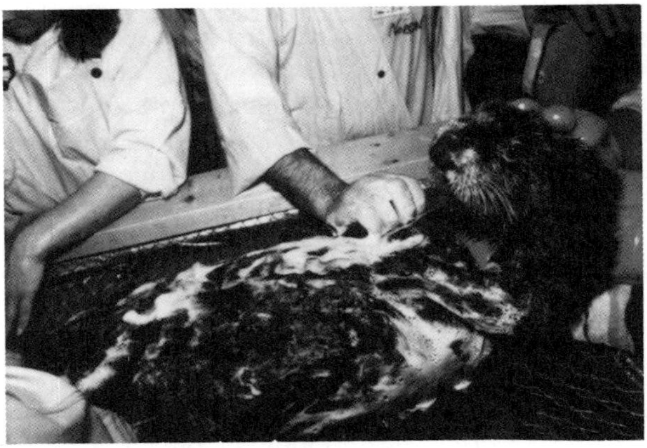

At the Valdez Otter Rescue center, a sedated otter is bathed with detergent to remove oil, then dried to restore the heat retaining insulation properties of its pelt. Photo by Tony Dawson.

in water. The fur immediately became saturated, whereupon the animal had to be removed, redried, and left overnight in a dry pen. It was released into the wild the following day and tracked for 24 hours before technical difficulties halted further observations.

Williams *et al.* (1988) made a concerted effort to develop a safe and effective procedure to clean and rehabilitate sea otters. After being fouled, mildly sedated otters were cleaned with 4–8 L of a 1 : 16 solution of Dawn detergent (Proctor and Gamble), which in a previous study had been shown to be the most effective of six cleansers tested. Dawn met two important criteria: it not only removed the contaminating oil, but if used correctly, it restored "loft" to the pelt, and thus the insulating air layer within the fur. Otters were washed for 40 minutes, then for 40 minutes rinsed with a shower head under moderate water pressure; air entrained in the water helped to restore the pelt. The animals were then dried with towels. Judging from observations and results of clinical and physiological tests, the investigators concluded that "sea otters that have had 20% of their body surface area oiled can be successfully cleaned and rehabilitated in 1–2 weeks . . . . However, normal grooming by the cleaned otter is essential for the full restoration of the fur. One or two additional weeks may be required for a complete recovery from the stress of oiling and cleaning . . . ."

The critical factor, it seems (one that had frustrated previous attempts), is the need to adequately rinse the detergent from the pelt. Otherwise, even small quantities act as a wetting agent and allow water to penetrate the fur. The investigators also pointed out the need to provide both water and haul-out space during recovery. These afford an animal the opportunity to enter the water for cooling and to leave it for warming. The choice is left to the animal. Otters in the study hauled out for various lengths of time after they had been cleaned. This study by Williams *et al.* (1988) served as the basis for rehabilitating sea otters fouled by the *Exxon Valdez*.

Three hundred and sixty-four oiled otters were processed at centers in Valdez, Seward, Homer, and Kodiak, Alaska; 53% were rehabilitated and returned to the wild, 32% died at the centers, and the rest were distributed to marine aquaria. The rescuers, including one of this chapter's authors (T. D. Williams), noted that certain steps were critical to success. They found that survival rate greatly improved when the otters were treated first with fluids, antibiotics, and vitamins, and allowed to rest. Attempts to feed them met with little success, and the small amount they did eat seemed to be poorly absorbed from their oil-coated intestines. Only after the otters were stabilized did the process of cleaning begin. First, they were sedated to diminish anxiety and protect the handlers from bites. After being bathed with detergent, the otters were dried for 1 hour, then placed in a cage for 12 hours before transfer to a saltwater bath. This was an improvement over freshwater, which had been used in prior attempts. Two animals placed together in a bath fared better than when alone.

The entire process was costly and involved a great deal of dedication and effort. The true measure of success will lie in the long-term health of the animals that were released. Radio transmitters were implanted in 45 of them to track their movements and thereby determine whether they can successfully re-establish themselves.

Summary

Oil poses the greatest threat to marine mammals with hair or fur; of these the otter is clearly the most vulnerable. The sea otter is small, energetic, and unable to tolerate interruptions in feeding. The dense fur coat, an ideal matrix for retaining oil, must be kept groomed, and therefore free of oil. An otter that might escape the consequences of ingesting oil while grooming a fouled coat might later be exposed to systemic effects of eating contaminated prey.

As a group, the vulnerability of sea otters is compounded by their behavior, which tends to fix them in an area whether or not it contains oil. And displacement, whether by their own response or through well-meaning intervention, does not necessarily guarantee survival.

References

Anonymous (1971). Alaska Peninsula oil spill. *Smithson. Inst., Cent. Short-Lived Phenom., Annu. Rep., 1970*. Event No. 36–70, pp. 154–157.

Anonymous (1980). Dead oiled sea otter. *Smithsonian Inst., Sci. Event Alert Network (SEAN) Bull.* **5**(11), 34.

Anonymous (1982). Dead sea otter with tar patch on fur. *Smithsonian Inst., Sci. Event Alert Network (SEAN) Bull.* **7,** 33.

Baker, J. R., Jones, A. M., Jones, T. P., and Watson, H. C. (1981). Otter *Lutra lutra* L. mortality and marine oil pollution. *Biol. Conserv.* **20,** 311–321.

Barabash-Nikiforov, I. I., Reshetkin, V. V., and Shidlovskaya, N. K. (1947). "The Sea Otter (Kalan)," Rep. No. OTS 61-31057. Natl. Tech. Inf. Serv., Springfield, Virginia. (transl. by Isr. Prog. Sci. Transl.).

Cimberg, R. L., and Costa, D. P. (1985). North Aleutian shelf sea otters and their vulnerability to oil. *In* "Proceedings of the 1985 Oil Spill Conference: Prevention, Behavior, Control, Cleanup," pp. 211–217. Am. Pet. Inst., Washington, D.C.

Costa, D. P., and Kooyman, G. L. (1982). Oxygen consumption, thermoregulation, and the effect of fur oiling and washing on the sea otter, *Enhydra lutris*. *Can. J. Zool.* **60,** 2761–2767.

Coues, E. (1877). Fur-bearing animals. Government Printing Office, Washington, D.C., pages 294–348. (Reprinted by Arno Press and the New York Times, 1970).

Davis, R. W., Williams, T. M., Thomas, J. A., Kastelein, R., and Cornell, L. H. (1988). The effects of oil contamination and cleaning on sea otters (*Enhydra lutris*). II. Metabolism, thermoregulation, and behavior. *Can. J. Zool.* **66,** 2782–2790.

Elliot, H. W. (1875). "A Report upon the Condition of Affairs in the Territory of Alaska." U.S. Govt. Printing Office, Washington, D.C.

Gentry, R. L., and Peterson, R. S. (1967). Underwater vision of the sea otter. *Nature* (*London*) **216,** 435–436.

Hess, R., and Trobaugh, L. (1971). Kodiak Island oil pollution. *Smithson. Inst., Cent. Short-Lived Phenom. Annu. Rep., 1970* Event, No. 26–70, pp. 150–153.

Howe, J. R. (1983). The sea otter puzzle. *Audubon* **85**(1), 34–39.

Irons, D. B., Nysewander, D. R., and Trapp, J. L. (1988). "Prince William Sound Sea Otter Distribution in Relation to Population Growth and Habitat Type." U.S. Fish Wildl. Serv., Anchorage, Alaska (unpublished report).

Jameson, G. L. (1986). "Trial Systematic Salvage of Beach-Cast Sea Otter, *Enhydra lutris*, Carcasses in the Central and Southern Portion of the Sea Otter Range in California," Rep. Contract No. MM2629849-8. U.S. Mar. Mammal Comm., Washington, D.C.

Kenyon, K. W. (1969). "The Sea Otter in the Eastern Pacific Ocean." Dover, New York.

Kooyman, G. L., Gentry, R. L., and McAllister, W. B. (1976). "Physiological Impact of Oil on Pinnipeds," Rep. Northwest Fish. Cent., Natl. Mar. Fish. Serv., Natl. Ocean. and Atmos. Admin., Seattle, Washington.

Kooyman, G. L., Davis, R. W., and Castellini, M. A. (1977). Thermal conductance of immersed pinniped and sea otter pelts before and after oiling with Prudhoe Bay crude. *In* "Fate and Effects of Petroleum Hydrocarbons in Marine Ecosystems and Organisms" (D. A. Wolfe, ed.), pp. 151–157. Pergamon, New York.

Limbaugh, C. (1961). Observations on the California sea otter. *J. Mammal.* **42,** 271–272.

McEwan, E. H., Aitchison, N., and Whitehead, P. E. (1974). Energy metabolism of oiled muskrats. *Can. J. Zool.* **52,** 1057–1062.

Nachtigall, P. E. (1986). Vision, audition, and chemoreception in dolphins and other marine mammals. *In* "Dolphin Cognition and Behavior: A Comparative Approach" (R. J. Schusterman, J. A. Thomas, and F. G. Wood, eds.), pp. 79–113. Erlbaum, Hillsdale, New Jersey.

Richardson, M. G. (1979). ESSO BERNICIA incident, Shetland, December/January. *Mar. Pollut. Bull.* **10**(4), 97.

Siniff, D. B., Williams, T. D., Johnson, A. M., and Garshelis, D. L. (1982). Experiments on the response of sea otters, *Enhydra lutris,* to oil contamination. *Biol. Conserv.* **23,** 261–272.

Van Blaricom, G. R., and Jameson, R. J. (1982). Lumber spill in central California waters: Implications for oil spills and sea otters. *Science* **215,** 1503–1505.

Williams, T. D. (1978). "Chemical Immobilization, Baseline Hematological Parameters and Oil Contamination in the Sea Otter," U.S. Mar. Mammal Comm. Rep. No. MMC-77/06. Natl. Tech. Inf. Serv., Springfield, Virginia.

Williams, T. M., Kastelein, R. A., Davis, R. W., and Thomas, J. A. (1988). The effects of oil contamination and cleaning on sea otters (*Enhydra lutris*). I. Thermoregulatory implications based on pelt studies. *Can. J. Zool.* **66,** 2776–2781.

Wragg, L. E. (1954). The effect of DDT and oil on muskrats. *Can. Field Nat.* **68,** 11–13.

9

Polar Bears and Oil: Ecologic Perspectives

Ian Stirling
Canadian Wildlife Service
and
Department of Zoology
University of Alberta
Edmonton, Alberta
TG6 2E9 Canada

Introduction

Bears are found on all the continents of the world except Australia and Antarctica, yet only one species, the polar bear, *Ursus maritimus*, evolved to take advantage of the marine ecosystem. The ice along the coast of the polar basin provided a platform upon which the ancestral polar bear could move from the safety of land onto the surface of the sea to hunt seals. This change in life style set the stage for the polar bear to diverge from its brown bear ancestors only a quarter of a million years or so ago. The polar bear quickly evolved its white coat, elongated skull, small ears, and large feet in response to the selection pressures of life in its new environment. Although its outward appearance is now quite different from its relatives, the fact that it readily interbreeds with brown bears in captivity, and that their offspring are also capable of interbreeding, indicates that the evolution of the polar bear is indeed a recent event. Unlike the seals or whales, polar bears have not evolved the ability to stay underwater for prolonged periods. Nor have they developed webbing between their toes like sea otters, or other morphological adaptations typical of pinnipeds or cetaceans. Yet polar bears spend as much of the year as possible on sea ice, and are completely dependent on the sea for their existence; they are true marine mammals in an ecologic sense.

Sea Mammals and Oil: Confronting the Risks
Copyright © 1990 by Academic Press, Inc.
All rights of reproduction in any form reserved.

Distribution

Polar bears presently range throughout the ice-covered waters of the Arctic (Fig. 9–1), and occur as local subpopulations of a few hundred to a few thousand. Their seasonal distribution and movements vary considerably between areas, depending on the local annual pattern of ice formation, distribution, and breakup. During the winter and spring, polar bears are distributed offshore from the mainland and various Arctic archipelagos. As the sea ice breaks up in the spring and early summer, bears follow the receding ice edge, which harbors their preferred prey, ringed seals and, to a lesser extent, bearded seals. In regions where the ice melts completely during the summer months, polar bears are forced onto the land and either walk to another area where the sea is still covered with ice, or wait for freeze-up, as they do along the shores of Hudson Bay. There, the bears must spend about 4 months on land living on stored fat. Toward the end of

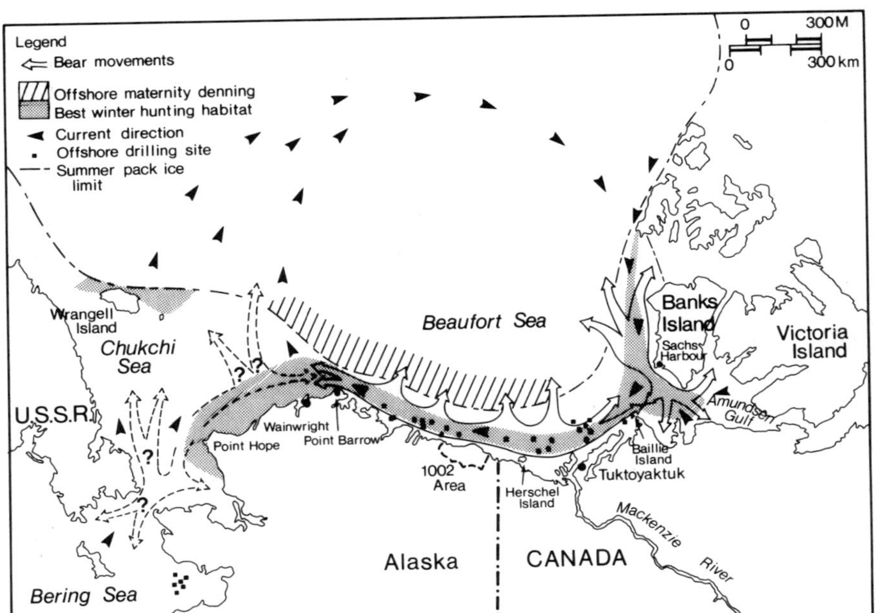

Figure 9-1
General distribution of prime polar bear habitat in the Beaufort, Bering, and Chuckchi seas so far as it is known; approximate distribution and seasonal movement patterns of different subpopulations of polar bears; a sample of offshore drilling sites; and the direction of major ocean currents.

this annual fast some bears, notably subadults and females with cubs, deplete most of their stored reserves, become hungry, and explore alternative food sources, such as the dump at Churchill (Lunn and Stirling, 1985).

Individual polar bears show a high degree of seasonal fidelity to particular areas within their home ranges, yet are not fixed to restricted locations. Unlike terrestrial bears which defend territories containing the best feeding areas, polar bears must move in response to annual variation in seal distribution. In good hunting areas, densities of polar bears may increase greatly as long as seals are abundant. More often, polar bears are widely distributed on the sea ice at fairly low densities as solitary animals or family groups, and, during the spring, as breeding pairs.

Life History

Polar bears mate in April and May. Following delayed implantation, the fertilized egg (blastocyst) begins to develop in September or October. In early November pregnant females dig maternity dens in snowdrifts on land near the coast, or offshore in the multi-year pack ice in the Beaufort Sea (Lentfer 1975; Amstrup, 1986). Cubs weighing less than a kilogram are born in December and early January. By the time they weigh 10–15 kg, they are able, with their mothers, to hunt seals on sea ice. In most areas of the Arctic, cubs remain with their mothers for 2½ years before they are weaned, although in southern Hudson Bay, some females wean their cubs at 1½ years. Consequently, the reproductive interval for most females is 3 years, or longer if they miss a pregnancy. This long period of weaning is necessary for cubs to learn hunting methods and seasonal movements. It also means that polar bears have a very low reproductive rate. Thus, populations recover slowly from declines caused by overhunting or environmental disruption. When fully grown, adult male polar bears weigh from 450–600 kg and females from 150–300 kg, although pregnant females can reach about 500 kg. Maximum recorded age in the wild is 32 years for females and 28 years for males, although few live longer than 20 years. Intense competition for mates results in wounds and broken canine teeth in males, contributing to their shorter life expectancy (Ramsay and Stirling, 1986).

Migration

Seasonal movements of polar bears in the Beaufort, Bering, and Chukchi seas are influenced primarily by the state of the sea ice and its effect on the distribution of prey. In summer, open water prevails along the coastlines and the polar bears move north to remain with the pack ice where they can still hunt seals (see Fig. 9–1); a few bears may be stranded on land. In fall, the polar pack begins to drift south again, and young ice forms between it and the land (Lindsay, 1975, 1977). The number of polar bears and the timing of their arrival along the coast is closely correlated with the density and southward advance of polar pack ice (Lentfer, 1972; Stirling, 1974a). During winter and early spring, bears congregate to hunt along the leads that parallel the coast (see Fig. 9–1). In the Beaufort Sea, individuals may move up to several hundred kilometers back and forth along these leads (Amstrup, 1986). Consequently, a significant portion of the local population is liable to pass near any particular point in the transition zone during the course of a winter.

During the late spring and summer in the High Arctic, polar bears, especially subadults, sometimes swim for several hours among the ice floes. Up to 4% of their time can be spent in the water. Tagged bears have been recorded swimming across bays and wide leads, and even larger bodies of water. In eastern Barrow Strait, they have been tracked for a hundred kilometers or so.

Habitat Preference

The most important single factor influencing the vulnerability of polar bears to oil is their habitat preference. Although polar bears occur throughout the Arctic, their distribution within that vast and variable habitat is determined by the distribution of their principal prey species, the ringed seal and the bearded seal, which in turn is largely influenced by ice conditions and water depth (Stirling *et al.*, 1982; Kingsley *et al.*, 1985).

Stirling *et al.* (1981) determined the habitat preference and distribution of polar bears in the eastern Beaufort Sea and Amundsen Gulf from sightings of bears and their tracks, along approximately 74,500 km of survey, from 1971 through 1979. They identified seven broad categories of sea ice habitat, of which only the following three were important to polar bears: (1) *stable fast ice* with deep snowdrifts along the pressure ridges suitable for ringed seal birth lairs (Smith and Stirling, 1975), usually occurring in mouths of bays and as land-fast ice out from coastlines; (2) the *floe edge* where leads are wide (>1 km), usually with small open and refrozen leads parallel to or emanating from it; and (3) areas of *moving ice* with seven-eighths or more cover, where wind and currents cause much movement of ice followed by refreezing that creates intermittent lanes or patches of refrozen young ice.

Adult male polar bears show a strong preference for the floe edge and moving ice over areas of extensive stable ice. Subadult males show similar though slightly less marked preferences. In contrast, adult females with first year cubs prefer areas of stable fast ice, probably because there are fewer bears here, especially adult males which might steal their killed seals or attack their cubs. The distribution of subadult females was similar to that of adult females with young cubs. Lone adult females and females with older cubs show no statistically significant preference for any of the three preferred habitat types, although more are caught along the floe edge and active ice areas.

The most likely reason for the greater number of polar bears in the floe edge and active ice habitats is that seals are more abundant here, and probably more accessible. Some knowledge of how seals are distributed helps us understand habitat preference and distribution of the bears. As freeze-up progresses during autumn, ringed and bearded seals maintain breathing holes by abrading young ice with their heavy foreflipper claws (Smith and Stirling, 1975). Breeding adult ringed seals tend to concentrate in areas of fast ice that remain frozen through most if not all of the winter. Here, pregnant females scoop out subnivean birth lairs in the deep drifts that develop over their breathing holes. The snow cover provides protection from predators. This protection is lacking along the floe edge or in the areas of active ice where there may be little or no snow cover. Seal pups in these areas are more vulnerable to predation.

The location of leads strongly influences the distribution of polar bears hunting during winter. This was dramatically demonstrated in a study survey conducted in March 1983 by D. Andriashek of the Canadian Wildlife Service. During a 2500-km search over completely frozen ice, only six bears were seen. After a week of strong winds opened a fresh lead about 150 km offshore, 90 bears were seen in a search of 3850 km in the next six days. Few tracks were seen outside a corridor of about 1 kilometer on either side of the lead, and no bears were sighted during approximately 2800 km of survey over sea ice away from the lead.

The most important floe edge and moving-ice habitats are distributed in a band of varying width which overlaps the shoreleads along the west coast of Banks Island, through the Bathurst Polynya at the entrance to Amundsen Gulf, and westward parallel to the north coast of Canada and Alaska (see Fig. 9–1) (Smith and Rigby, 1981). This intermediate zone of fractured, unconsolidated, annual ice lies over relatively shallow water between the annual landfast ice along the coast and the multiyear polar pack ice further offshore. In Canada, it is usually referred to as the transition zone, and in the United States as the seasonal pack ice zone (Dickens *et al.*, 1987). Along the mainland coast of the eastern Beaufort Sea, the location of the seaward edge of landfast ice is generally predictable; it roughly follows the 20-m depth contour (Cooper, 1974). Its position in the western Beaufort Sea is similar. The most important stable ice habitat in most areas tends to lie between the coastline and the floe edge and in deep bays.

Exploration and Production Activities in Prime Habitats

In the eastern Beaufort Sea, offshore drilling has taken place along the mainland coast north of the Mackenzie Delta and Tuktoyaktuk Peninsula. Because of technical limitations on equipment available (Croasdale, 1986), most of the activity has been in shallower areas (<50 m). Thus, there is overlap between the most important polar bear feeding habitat and offshore drilling activities (Stirling, 1988).

Large-scale spills or blowouts from these wells could affect prime habitat for polar bears in the Beaufort Sea. The problem would be magnified by a predominant east-to-west current (Wilson, 1974) that would carry oil into the transition zone in the floe edge and moving ice habitats. In fall, oil incorporated into newly forming ice would be similarly transported, but not released until spring (Neff, this volume, Chapter 1). Ice in the Beaufort Sea moves from east to west at speeds that vary from about 4 km/day in March and April to 15 km/day in September and October (Melling, personal communication). Consequently, oil entrapped in ice would also reach the floe edge and moving-ice habitats in the transition zone where polar bears of all age classes and both sexes are most abundant.

Oil production activities are also situated in the path of the polar bears' annual migration across the southern Beaufort Sea (see Fig. 9–1). Thus, bears might move through a polluted area. The number contaminated would vary with season, location, and the number of leads. In most years, migration along leads peaks in spring and fall.

Spills from tankers and, to a lesser degree, support vessels represent another threat, particularly since their routes could pass through prime polar bear habitat in the Beaufort Sea. For example, oil from wells in the Mackenzie Delta northwest of Tuktoyaktuk could be shuttled to a supertanker stationed west of Point Barrow, Alaska. To do so would require several trips by a smaller ice-reinforced tanker. The supertanker would then head for southern ports via the Chukchi and Bering seas. It is clear that the risk to polar bears will increase with the extent and frequency of such activities.

Habitat Modification

In recent years ingenious techniques have been developed to extend offshore drilling long after freeze-up and, in some cases, through the entire winter (Croasdale, 1986; Stirling, 1988). These include artificial islands, ice islands, the floating Conical Drilling Unit, the Single Steel Drilling Caisson (which is ballasted to a subsea berm) and, possibly most relevant of all to this discussion, the Mobile Arctic Caisson (called the Molikpaq) (Fig. 9–2). The Molikpaq was designed for drilling to depths of up to 6000 m in 20–40 m of water and appears to be the prototype for future development. It is intended to operate in a depth range that overlaps much of the best floe edge and active ice habitat in the transition zone.

Figure 9-2

The Molikpaq drilling offshore in the Beaufort Sea, northeast of Tuktoyaktuk in March. Note the cracks in the ice formed in the lee of the drilling platform as the annual ice drifts past (photo by Ransom Photographers, for Gulf Canada).

Polar bears have been observed in the vicinity of the Molikpaq throughout the winter. Most bears do not remain in the area for long, but under certain circumstances, platforms and artificial islands can influence their distribution and movement. During periods of calm, cold weather, particularly in late March and early April when sea ice is thickest, most of the cracks where polar bears can hunt freeze over. However, the current-induced movement of ice from east to west causes leads to form "down-stream" from the installations (Fig. 9–2). These leads are used routinely by small numbers of seals, though as many as 30 or 40 have been observed at one time. Polar bears, including adult females and their cubs, are attracted to these leads when others are closed (Stirling, 1988). Some bears stay for several days and possibly even for a few weeks, eventually dispersing as other leads open later in spring. In the event of a spill or blowout from a well, it is likely that bears hunting around the rigs would be the first to be oiled.

Behavior

Hunting

The hunting strategy of a polar bear can promote contact with oil. The hunting animal typically lies motionless beside an open or snow-covered breathing hole, or at the edge of a lead, and strikes at seals that surface here. In doing so, the bear reaches into the water with its jaws, often submerging its head and occasionally diving into the lead. Another strategy, used during late spring and summer after the snow has melted, is to stalk basking seals by swimming along the edge of the ice. A bear may dive and swim underwater between natural holes in the ice, surfacing quietly to breathe, until it arrives next to its prey, at which point it suddenly surfaces to attack. Bears also obtain food by scavenging remains left by other bears.

Polar bears spend more time in hunting than in any other activity. During the spring, adult females with cubs and adult males will spend 19–25% of their time hunting, and in the summer 30– 50% (Stirling, 1974b; Stirling and Latour, 1978). Hunting success varies seasonally and among individuals. In early summer—the best time for hunting seals—adult females with cubs catch one seal every 4–5 days. During winter, the interval between successful catches is much longer, judging from the fact that bears of all ages and both sexes are lighter in weight in March or April than in July. Because seals are less available in winter, bears may explore alternative foods, leading them more frequently to fatal encounters with humans at that time of year.

Reproduction

To a degree, polar bear reproductive behaviors isolate them from contact with oil. From mid-April to late May, breeding pairs range along the floe edge and active ice areas that predominate along the transition zone and around polynyas such as those near Cape Bathurst or Point Barrow. Males accompany females for a week or more, apparently to induce ovulation, and may lead them away from areas where competing males are abundant (Ramsay and Stirling, 1986). Thus, for short periods, breeding polar bears may not be near areas where risk of contact with oil might be greatest.

In fall, pregnant females move away from areas frequented by other bears. In the Beaufort Sea region, most maternity denning on land occurs on the west and south coasts of Banks Island. Few dens have been reported in recent years along the mainland coast of Alaska and Canada, possibly because excessive hunting pressure prior to the early 1960s eliminated those individuals showing some fidelity to the area (Stirling *et al.*, 1975). Amstrup (1986) has shown that 80% or more of maternity denning along the mainland coast of northern Alaska and Canada takes place in multiyear pack ice up to about 300 km offshore. Bears in those dens are likely too far out to contact oil during the winter.

The risk of exposure would be greater for polar bears in dens along the mainland coast in the vicinity of Herschel Island and along the Alaskan North Slope. After leaving the dens in spring, females evidently hunt seal pups in stable fast ice on the landward side of the transition zone. Eventually, they must cross it to reach the retreating pack ice where they spend the summer. A lead fouled by a spill might obstruct their path or contaminate bears migrating through it.

Grooming

Polar bears are clean animals. Over the years, we have caught several thousand, very few of them greasy or dirty. Grooming is necessary to maintain thermal efficiency of their pelt, which is often soiled after feeding. Polar bears feed on seals by tearing pieces of skin and blubber away from the carcass; they take fat preferentially. After feeding for 20–30 minutes, a bear will wash its paws and face in a nearby lead or melt pond on the ice (Stirling, 1974b). A bear may repeat this behavior several times while eating, and again after it is done. During colder weather when water is not available, a bear will clean itself by licking its paws and rolling or pushing its face and neck against the snow. Mutual grooming has also been observed among females and cubs of all ages. Thus, it is certain that polar bears will attempt to remove any foreign substance, including oil, from their fur.

Investigative Behavior

To survive in an extreme environment, such as the Arctic sea ice, a predator must be opportunistic and investigative. It is not surprising that polar bears are extremely curious and will explore anything novel in their environment. They are also scavengers, and will eat an extraordinary variety of unnatural foods, some so obviously lacking in nutrition that the reason for eating them is unclear. For example, styrofoam, pieces of plastic, and even parts of a car battery (which apparently killed the animal) have all been found in the stomachs or scats of polar bears (Russell, 1975; Lunn and Stirling, 1985). Similarly, Inuk hunters have observed polar bears biting cans of snowmobile oil, and there are instances of bears biting and causing leaks in neoprene fuel bladders. A polar bear in Alaska died after drinking a mixture of rhodamine B dye and ethylene glycol that was being used to mark runways in the winter (Amstrup *et al.*, 1989). These anecdotes raise a number of questions. Does the smell of refined or crude oil attract polar bears? Would one scavenge a dead oil-covered seabird or seal, thereby ingesting toxic materials? Might a bear actively avoid an oil-covered lead if it could, or would it simply swim through the oil?

Summarizing the Risk

The ways in which a bear gathers food, the characteristics of its environment, and its migratory behavior all influence its chances of contacting spilled oil. The polar bear is closely associated with ice floes, the same type of environment which entrains, concentrates, and moves surface oil, and retards its degradation and elimination from the ecosystem. The polar bear shares with the sea otter the need to maintain a pristine hair coat as protection against heat loss. Consequently, exposure to oil at any stage of its life cycle is apt to be harmful or even fatal.

Acknowledgments

I acknowledge the Canadian Wildlife Service, Polar Continental Shelf Project, National Science and Engineering Research Council, the Department of Zoology of the University of Alberta, and the World Wildlife Fund (Canada) for their continued support of my research on polar bears. I would also like to thank H. Melling for permission to cite his unpublished data, and the following for their invaluable assistance in the laboratory and field and for helping to broaden my understanding of polar bears: D. Andriashek, W. Calvert, A. Derocher, P. B. Latour, M. A. Ramsay, and T. G. Smith.

References

Amstrup, S. C. (1986). Research on polar bears in Alaska, 1983–85. *In* "Proceedings of the 9th Working Meeting of the IUCN Polar Bear Specialist Group, Edmonton, Alberta 1985," pp. 85–112. Int. Union Conserv. Nature and Nat. Resour. Publ., Gland, Switzerland.

Amstrup, S. C., Gardner, C., Myers, K. C., and Oehme, F. W. (1989). Ethylene glycol (antifreeze) poisoning in a free-ranging polar bear. *Vet. Hum. Toxicol.* **31,** 317–319.

Cooper, P. F., Jr. (1974). Landfast ice in the southeastern part of the Beaufort Sea. *In* "The Coast and Shelf of the Beaufort Sea" (J. C. Reed and J. E. Sater, eds.), pp. 235–242. Arct. Inst. North Am., Arlington, Virginia.

Croasdale, K. R. (1986). Arctic offshore technology and its relevance to the Antarctic. *In* "Antarctic Treaty System, An Assessment" (Polar Research Board, ed.), pp. 245–264. Natl. Acad. Press, Washington, D.C.

DeMaster, D. P., and Stirling, I. (1981). *Ursus maritimus. Mamm. Species* **45,** 1–7.

Dickens, D., Martin, L., Bjerkelund, I., Potter, S., Erickson, D., Harper, J., Norton, P., Johnson, S., and Vonk, P. (1987). "Environmental Atlas for Beaufort Sea Oil Spill Response." D. Dickens Assoc. Ltd., Vancouver, B.C.

Kingsley, M. S. K., Stirling, I., and Calvert, W. (1985). The distribution and abundance of seals in the Canadian High Arctic, 1980–82. *Can. J. Fish. Aquat. Sci.* **42,** 1189–1210.

Lentfer, J. W. (1972). Polar bear—sea ice relationships. *In* "Bears—Their Biology and Management" (S. Herrero, ed.), IUCN New Ser. No. 23, pp. 165–171.

Lentfer, J. W. (1975). Polar bear denning on drifting sea ice. *J. Mammal.* **56,** 716.

Lindsay, D. G. (1975). "Sea Ice Atlas of Arctic Canada, 1961–1968," Rep. Dept. of Energy, Mines and Resources, Ottawa, Canada.

Lindsay, D. G. (1977). "Sea Ice Atlas of Arctic Canada, 1969–1974," Rep. Dept. of Energy, Mines and Resources, Ottawa, Canada.

Lunn, N. J., and Stirling, I. (1985). The ecological significance of supplemental food to polar bears during the ice-free period of western Hudson Bay. *Can. J. Zool.* **63,** 2291–2297.

Ramsay, M. A., and Stirling, I. (1986). On the mating system of polar bears. *Can. J. Zool.* **64,** 2142–2151.

Russell, R. H. (1975). The food habits of polar bears of James Bay and Southwest Hudson Bay in summer and autumn. *Arctic* **28,** 117–129.

Smith, M., and Rigby, B. (1981). Distribution of polynyas in the Canadian Arctic. *Can. Wildl. Serv., Occas. Pap.* **45,** 7–28.

Smith, T. G., and Stirling, I. (1975). The breeding habitat of the ringed seal (*Phoca hispida*): The birth lair and associated structures. *Can. J. Zool.* **53,** 1297–1305.

Stirling, I. (1974a). Polar bear research in the Beaufort Sea. *In* "The Coast and Shelf of the Beaufort Sea" (J. C. Reed and J. E. Slater, eds.), pp. 721–733. Arct. Inst. North Am., Arlington, Virginia.

Stirling, I. (1974b). Midsummer observations on the behavior of wild polar bears (*Ursus maritimus*). *Can. J. Zool.* **52,** 1191–1198.

Stirling, I. (1988). Attraction of polar bears to offshore drilling sites in the eastern Beaufort Sea. *Polar Rec.* **24:**1–8.

Stirling, I., and Latour, P. B. (1978). Comparative hunting abilities of polar bear cubs of different ages. *Can. J. Zool.* **56,** 1768–1772.

Stirling, I., Andriashek, D., Latour, P., and Calvert, W. (1975). "The Distribution and Abundance of Polar Bears in the Eastern Beaufort Sea," Final Report to the Beaufort Sea Project. Fish. Mar. Serv., Dep. of the Environment, Victoria, B.C.

Stirling, I., Andriashek, D., and Calvert, W. (1981). "Habitat Preferences and Distribution of Polar Bears in the Western Canadian Arctic," Rep. Can. Wildl. Serv., Edmonton, to Dome Petroleum, Calgary, Alberta.

Stirling, I., Kingsley, M. C. S., and Calvert, W. (1982). The distribution and abundance of seals in the eastern Beaufort Sea, 1974–79. *Can. Wildl. Serv., Occas. Pap.* **47,** 1–23.

Wilson, H. P. (1974). Wind and currents in the Beaufort Sea. *In* "The Coast and Shelf of the Beaufort Sea" (J. C. Reed and J. E. Sater, eds.), Arct. Inst. North Am., Arlington, Virginia.

10

Physiologic and Toxic Effects on Polar Bears

David J. St. Aubin
Department of Pathology
Ontario Veterinary College
University of Guelph
Guelph, Ontario
N1G 2W1 Canada

Historical Notes

There have been few major spills in the polar bear's (*Ursus maritimus*) natural range, and no reports of bears critically fouled by oil. Anecdotal accounts show that bears occasionally come into contact with it. Stirling (this volume, Chapter 9) has learned of bears biting cans of motor oil and fuel storage bladders without evidence of immediate harm. Adding to these scant data are findings from a single laboratory study (Øritsland *et al.*, 1981). The effects of oil on bears can be predicted more from what we know about their physiology and life history than from these limited observations.

Detection and Avoidance

Stirling (this volume, Chapter 9) has vividly described the ways in which a polar bear gathers food. It hunts prey by keen eyesight and by detecting the odor of a ringed seal in a subnivean birth layer. These well-developed senses of sight and smell are common among carnivores and would seem to provide a bear with the means to detect spilled oil.

Sea Mammals and Oil: Confronting the Risks
Copyright © 1990 by Academic Press, Inc.
All rights of reproduction in any form reserved.

Some indication of their ability to do so can be gleaned from the laboratory study on three captive bears (Oritsland *et al.,* 1981). The animals were coaxed individually along a passageway leading to a small pool containing 7000 L of seawater covered with a 1-cm surface slick of crude oil. None entered the pool voluntarily, but all three investigated its oil-covered surface. As the bears stretched out over the pool to reach seal blubber suspended inaccessibly from the top of the cage, the door was closed behind them, forcing them into the water. The bears made deliberate attempts to escape and were able to do so to some extent by supporting themselves on the cage bars which encircled the pool. They continued trying to obtain the bait. When the cage door was opened after 15–50 minutes, the bears left the pool immediately. These brief observations make it clear that bears can detect oil and seem determined to avoid it. Their reaction to it in confinement offers a clue to how a free-ranging bear might behave in a wild setting.

The bear's annual cycle includes seasonal migrations across open leads as it follows the edge of the pack ice. Confronted by an oil-covered lead, it might investigate and avoid the contaminated water. But would it nevertheless enter the lead, bent on continuing its migration through an area that offered no clean alternative? Would it resist the lure of a seal, perhaps disoriented, emerging in the slick? In the captive setting, it was the desire to obtain food that caused the bears to lean precariously over the oiled pool. Their appetites were enhanced by food deprivation during the pre-experimental period, yet natural cycles in prey abundance could induce similar hunger in free-ranging bears. Ready access to seals confined to limited open water areas during the winter might in fact draw polar bears to contaminated areas.

Surface Contact

To protect them from the chill of their environment, polar bears rely principally on a thick coat of hair and subcutaneous fat (Schölander *et al.,* 1950; Øritsland, 1970). The effect of fouling on thermoregulation in these animals is thus a major concern. In an incomplete study, Øritsland *et al.* (1981) noted that thermal conductance through wet, oiled pelts was five to six times more rapid than through dry pelts. No meaningful conclusions can be drawn because no wet pelts were used as controls.

Nevertheless, it is reasonable to assume from studies of other furbearers (this volume, Chapters 4 and 8) that surface fouling will have a significant impact on energy metabolism of polar bears. In an attempt to quantify the effect, live bears were exposed to oil (Øritsland *et al.,* 1981).

The bears had been captive for 3–4 months before being placed in an oil-covered pool. Basal metabolic rate, determined 5–8 days after surgical implantation of temperature transmitters, was somewhat higher than expected, then rose tenfold when the animals exercised (Hurst *et al.*, 1982). Rigorous exposure to wind at −16 to −24°C caused skin temperatures to fall as much as 7°C, but had little effect on core temperatures. After the bears had been in the oiled pool for 15–50 minutes, they were again exposed to alternating periods of wind and calm air. The core temperature in two showed no change, whereas that in the third decreased steadily when the bear was exposed to wind, and rose again to pre-oiling values when the air was calm. One day after oiling, the metabolic rate of the bears had increased 27–86%. These studies confirm the assumption that an animal that depends on its coat for insulation will be metabolically affected by oiling (Kooyman *et al.*, 1977). However, the magnitude of oil impact on polar bears cannot be deduced from the Hurst *et al.* study, because of the inexplicably high metabolic rates observed prior to oil exposure.

Oil Ingestion

Stirling's (this volume, Chapter 9) description of grooming behavior in polar bears sets the stage for understanding how the problem of oil-coating can be transformed into one of ingestion. The experimental study (Øritsland *et al.*, 1981) confirmed the association. During the first few hours after oiling, all three bears groomed their paws and forelegs intensively. One rubbed its coat along the cage bars, while another attempted to use snow that was provided. The snow became fouled and the bears consumed some of it. Grooming activity subsided over the next 5 days, though the bears were still covered with oil. Ingested oil caused vomiting and diarrhea. Hydrocarbons were absorbed into the circulation, distributed to various tissues, and excreted by bile and urine. During the 4 weeks after oiling, there was biochemical evidence that the bears were developing liver and kidney failure, and a disorder of red blood cell formation. Twenty–six days after oiling, one of the bears died, and 3 days later, another was euthanized.

The principal findings at necropsy were degeneration of kidney tubules, low-grade liver lesions, depressed lymphoid activity, and fungus-containing ulcers in the gastrointestinal tract. This suite of pathologic changes suggests that the toxic effects of oil were compounded by the stresses of the experiment. The bears were confined for metabolic studies, their diet was inadequate, drinking water was limited for a critical time after oiling, and surgical incisions and injection sites had become infected.

Lymphoid suppression and widespread mycotic lesions draw attention to the degree to which the animals were stressed. Could the stress of oiling alone produce these kinds of changes in a free-ranging bear? Perhaps so, when seals are scarce, temperatures are extreme, and depot fat stores are reduced.

Summary

The dramatic consequences of oil exposure, albeit intensified by the experimental protocol of Øritsland *et al.* (1981), highlight the vulnerability of polar bears. There is no reason to expect that they would escape harm should they become oiled in the wild. Imagine, if you will, a polar bear on the edge of a lead fouled with oil. A single plunge into it while chasing a seal, or emerging through it on return, is likely to affect the bear in a way comparable to a sea otter (this volume, Chapter 8), forcing it to groom compulsively at the expense of other vital activities. In a futile effort to restore the insulative quality of its coat, the bear ingests oil, adding to its metabolic stress.

More than any other marine mammal, bears scavenge food and are likely to manipulate and consume an oil-soaked seal or bird carcass. Judging from anecdotes of their behavior in the wild, they show no particular aversion to the taste of oil. Witness also the reaction of two of the study bears when placed in a holding cage previously occupied by an oil-fouled bear. Both became preoccupied with cleaning their lightly soiled paws, then turned their attention to the cage itself, actively licking oil residues from the bars and floor. For polar bears, the problems associated with ingesting fresh oil seem to far outweigh the risks of consuming trace amounts of petroleum hydrocarbons in the tissues of their prey.

An important lesson from the oiling study (Øritsland *et al.*, 1981) is that stress may intensify effects of oil exposure. A healthy 500-kg bear might conceivably ingest a liter or more of oil without harm, yet the same quantity might create serious problems for an animal facing low food availability or stressed by unusual environmental conditions. Seasonal and annual fluctuations in prey abundance alter the body condition of free-ranging bears. A hungry bear is more likely to scavenge contaminated prey and to suffer the consequences of oil ingestion.

References

Hurst, R. J., Øritsland, N. A., and Watts, P. D. (1982). Body mass, temperature and cost of walking in polar bears. *Acta Physiol. Scand.* **115,** 391–395.

Kooyman, G. L., Davis, R. W., and Castellini, M. A. (1977). Thermal conductance of immersed pinniped and sea otter pelts before and after oiling with Prudhoe Bay crude. *In* "Fate and Effects of Petroleum Hydrocarbons in Marine Ecosystems and Organisms" (D. A. Wolfe, ed.), pp. 151–157. Pergamon, New York.

Øritsland, N. A. (1970). Temperature regulation of the polar bear (*Thalarctos maritimus*). *Comp. Biochem. Physiol.* **37,** 225–233.

Øritsland, N. A., Engelhardt, F. R., Juck, F. A., Hurst, R. A., and Watts, P. D. (1981). "Effects of Crude Oil on Polar Bears," Environ. Stud. Rep. No. 24. Northern Affairs Program, Dept. of Indian Affairs and Northern Development, Ottawa, Ontario, Canada.

Schölander, P. F., Walters, V., Hock, R., and Irving, L. (1950). Heat regulation in some arctic and tropical mammals and birds. *Biol. Bull.* (*Woods Hole, Mass.*) **99,** 225–236.

11

Oil Effects on Manatees: Evaluating the Risks

David J. St. Aubin and Valerie Lounsbury
Department of Pathology
Ontario Veterinary College
University of Guelph
Guelph, Ontario
N1G 2W1 Canada

Introduction

The order Sirenia contains two families: the Trichechidae (manatees), comprising three species of the genus *Trichechus,* and the Dugongidae (dugongs), represented by a single species. Their ancestral relationships are unclear, though likely they are descended from a terrestrial herbivore (Domning, 1982). Their closest living terrestrial relatives are elephants and hyraxes. Unique anatomical and physiological traits distinguish the sirenians from other marine mammals, as does their herbivorous diet. Like cetaceans, they are wholly aquatic, though they may lift their forebody from water to browse on shoreline vegetation.

The present discussion will focus on the West Indian manatee, the only sirenian in North American waters. Observations on the Amazonian manatee, African manatee, and dugong will be included to fill gaps in data or to provide important comparative information. Detailed accounts of manatee ecology and life history, beyond the scope of this review, have been compiled by Bertram and Bertram (1973), Hartman (1979), and Caldwell and Caldwell (1985).

Sea Mammals and Oil: Confronting the Risks
Copyright © 1990 by Academic Press, Inc.
All rights of reproduction in any form reserved.

Distribution and Abundance

In the broadest sense, the West Indian manatee ranges from Virginia to central Brazil, including the shores of the Gulf of Mexico and the islands of the Caribbean Sea (Fig. 11–1). Its distribution is discontinuous, and mainly in four principal locations: the Atlantic and Gulf coasts of Florida; the Yucatan Peninsula south to Honduras; between the Orinoco and Amazon Rivers; and the Greater Antilles, including Cuba, Puerto Rico, Hispaniola, and Jamaica. These populations are considered to be more or less distinct; seasonal movements and occasional strays provide only limited opportunities for mixing and recruitment.

The population of West Indian manatees is estimated at 8000 to 17,000 animals, most of which live in Guyana (Food and Agriculture

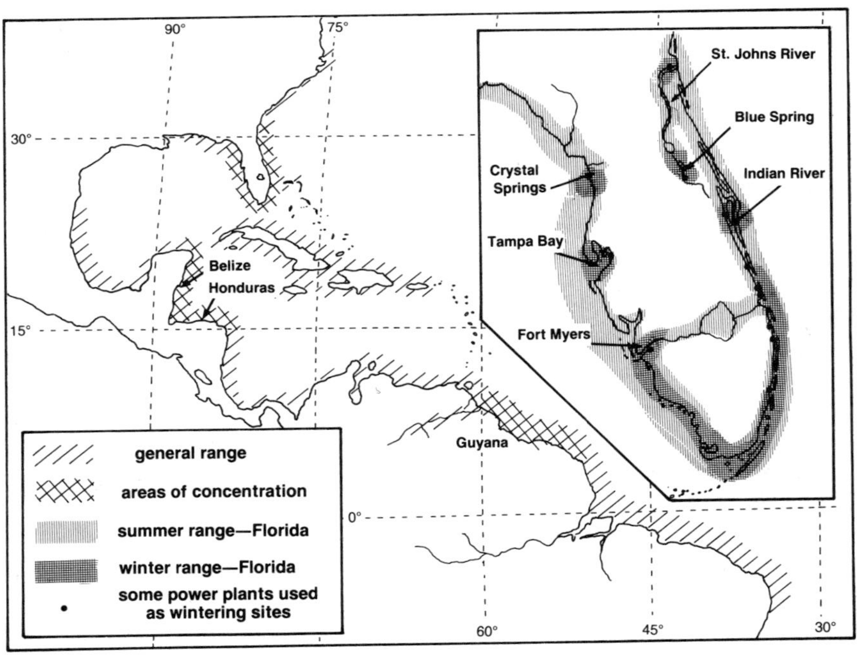

Figure 11-1
Distribution of the West Indian manatee. Favored wintering sites in Florida waters and the location of some power plants that attract manatees are indicated.

Organization, 1976). In Florida waters, estimates vary between 800 and 1200; most reports set the population at 1000 (Brownell *et al.,* 1978). In view of its restricted range and low numbers, the West Indian manatee has been listed by the International Union for Conservation of Nature and Natural Resources as vulnerable to extinction. It has been reduced from historical levels of several thousands by exploitation for its highly palatable meat (Bertram and Bertram, 1973) and, more recently, by accidental collisions with boats (Irvine *et al.,* 1978; Hartman, 1979; Odell and Reynolds, 1979).

Life History

Manatees breed throughout the year in Florida, although analysis of calf mortality indicates that low peaks in reproduction occur in spring and fall (Brownell *et al.,* 1978). Only the Amazonian manatee appears to be a truly seasonal breeder, having reproductive peaks that coincide with annual cycles in water level (Best, 1982). Copulation is preceded by vigorous courtship activity, belying the manatee's reputation for sluggish behavior. Bulls may escort a cow for a month or more, then engage in frantic pursuit through mud and vegetation before mating (Hartman, 1979).

Gestation lasts roughly a year, and cows apparently seek sheltered waters to give birth to their single calf (Hartman, 1979). The young may nurse for a year or more, but may begin grazing at the age of 2–3 months (Hartman, 1979). The association between cow and calf lasts up to 18 months (Powell and Waldron, 1978); minimum reproductive interval is therefore 2–2.5 years (Hartman, 1979), though some suggest that 3–4 (Powell and Waldron, 1978) or even 5 years (Brownell *et al.,* 1978) is more common.

Estimates of age at sexual maturity for female manatees vary from 3–5 years (Hartman, 1979) to 7–9 years (Odell, 1977); males mature at 9–10 years (Odell, 1977). Though there are no reliable estimates for average lifespan in the wild, captive animals have been maintained for 25–30 years (Brownell *et al.,* 1978; Hartman, 1979). By contrast, dugongs mature later and have a life expectancy of 50–60 years (Marsh *et al.,* 1984).

A newborn manatee is 1 m long and weighs 20–30 kg. Adults range up to 4 m in length and can weigh as much as 900 kg in weight. The Amazonian manatee is a much smaller animal; the largest on record is only 2.8 m long.

Preferred Habitat, Seasonal Movements, and Migration

Manatees dwell in protected, low salinity waters where vegetation is abundant. They range extensively into freshwater systems, including the St. Johns River in northern Florida, where they may congregate over 250 km from the sea (Powell and Waldron, 1978). As a rule they only venture into open oceanic waters to move from one favorable feeding area to another (Hartman, 1979). Such movements are generally confined to inshore waters less than 5 m deep, though extensive shallows off Florida's west coast can force animals many kilometers from land. Rare offshore sightings near the Dry Tortugas are thought to represent sporadic movements from other regional concentrations in the Caribbean Islands or even Mexico (Reynolds and Ferguson, 1984).

Seasonal movements are governed by the manatee's intolerance to cold. Historically, manatees used to retreat during the winter months to the southernmost areas of Florida, or sought natural warm springs such as those in Citrus County or the upper reaches of the St. Johns River (Campbell and Irvine, 1978). The advent of electric power plants which discharge warm water provided manatees with new alternative wintering sites. Their preference for discharge waters does not limit their movements entirely. Judging from the growth of marine algae and barnacles on the skin of animals returning to warm springs (Hartman, 1979), they also undertake lengthy excursions into the cold sea during winter.

With the return of warm weather, manatees disperse along the coast. Some return to the same summer feeding grounds (Shane, 1984), though their attraction to a specific area is not particularly strong (Hartman, 1979). In general, their movements during summer are unpredictable; they may range along several hundred kilometers of coastline as they explore alternative feeding areas (Hartman, 1979). Individuals may remain in the same area for several months or simply move on after only a few days.

Habitat Use and Physiology

Sirenian behavior and distribution are influenced by two important aspects of their physiology: nutrition and metabolism. As the only herbivorous marine mammals, they occupy a unique ecologic niche. Their distribution is limited to low-energy, inshore habitats supporting the growth of seagrasses. Their attraction to springs and rivers seems to be less one for freshwater than for specific types of vegetation.

Manatees consume a wide variety of aquatic and semi-aquatic plants, and show a mild preference for submerged succulent vegetation (Reynolds, 1978; Hartman, 1979). Their selection appears to change seasonally with availability of specific foods (Hartman, 1979). When preferred underwater types are depleted, they turn to floating mats of vegetation, algae, roots, or detritus, and at times will rise partly from the water to consume overhanging vegetation (Reynolds, 1978, 1981; Powell and Waldron, 1978; Hartman, 1979). While grazing they ingest considerable quantities of encrusting organisms such as diatoms, molluscs, and crustaceans (Best, 1981). They may take in foreign objects, such as plastic bags, pieces of rope and cloth, and artificial sponges (Reynolds, 1980).

Manatees must feed for 6–8 hours each day (Hartman, 1979) to compensate for the low energy content of their diet and for their low digestive efficiency, which in the Amazonian manatee ranges from 44–68% (Best, 1981). On average, manatees consume 8–13% of their body weight daily. They are not ruminants. To digest cellulose, they rely instead on hindgut fermentation, aided by secretions from a unique system of glands attached to the stomach (Reynolds, 1980). Digestion is slow; the transit time of food through the gut in the Amazonian manatee ranges up to 140 hours (Best, 1981).

The manatee's low-yield diet and quiet behavior match its extraordinarily low metabolic rate. Schölander and Irving (1941) first demonstrated that the rate is far below that predicted on the basis of body size. Subsequent studies on the West Indian and the Amazonian manatees confirmed that the rate is only 15–35% that of other mammals of comparable size (Gallivan and Best, 1981; Irvine, 1983; Gallivan *et al.*, 1983).

Slow metabolism accounts in part for the manatee's tropical range. Its ability to tolerate cold water is limited also by the high thermal conductance of its skin, which has only sparse sensory hairs and little blubber (Irvine, 1983). Only the Amazonian manatee develops seasonal reserves of subcutaneous fat, not for thermoregulation, but to offset annual fluctuations in food abundance (Gallivan *et al.*, 1983). In water below 20°C, manatees may increase their metabolic rate (Hartman, 1979) and allow core temperature to decrease, in effect decreasing the temperature gradient between the body and the environment (Gallivan *et al.*, 1983). These strategies provide limited benefit. Unseasonably cold temperatures, such as occurred during the winter of 1976–1977 in Florida, were responsible for increased mortality in the more northerly wintering sites (Campbell and Irvine, 1978).

Historical Notes on the Effects of Oil

Although information on the effects of oil on other marine mammal groups is patchy, it is voluminous when compared with the information available for sirenians. There has been no experimental study, and only a handful of observations suggesting that oil has harmed any sirenian.

No major spill has occurred in Florida waters occupied by manatees. To glean some understanding of the effects of oil, we can draw on observations from the Persian Gulf, where an apparent increase in dugong mortality coincided with the uncontrolled release of large quantities of oil from platforms damaged during the Iran–Iraq war. Between February and August, 1983, an estimated 30 million gal of crude oil spilled from eight wells in offshore Iranian oil fields (Miller, 1983). Continued fighting interfered with attempts to cap the wells or clean up the oil. An estimated 40–50% of the oil was presumed to have evaporated, leaving a mat of heavier fractions suspended just under the surface (Miller, 1983). By July, 53 dugong carcasses had been recovered in addition to an "unusual number" of fish, turtles, dolphins, snakes, and birds (Anonymous, 1983). So little was known of the dugong population in the area that the loss of at least 50 was thought to represent a major proportion of the local stock. The findings are difficult to interpret in the absence of baseline studies. In addition, detailed examinations were not performed in 1983, and we are left with an association between oil and mortality as ephemeral as the vapors that dissipated into the atmosphere.

Closer to home the relationship between oil and sirenian mortality is as obscure. During the winter of 1981 and 1982, three dead manatees were recovered near Jacksonville, Florida, and one in the Little Manatee River which empties into Tampa Bay (Table 11–1). The decomposed carcasses had tarlike material in the lower digestive tract (Anonymous, 1981a,b,c, 1982). Additional findings included plastic in the stomach of one animal (Anonymous, 1981c), and broken ribs associated with a superficial wound on the back of another (Anonymous, 1981a). No laboratory analyses were performed to confirm the composition of the substance presumed to be petroleum. It would be frivolous to imply an association between oil and the deaths of these manatees.

Presumed Effects of Oil

With no detailed information, the analysis of potential risks to manatees from oil is limited to speculation based on knowledge of their life history and ecology. First, how would manatees react to oil? Would they be able

Table 11–1
Reports of Manatees and Dugongs Associated with Oil

Date	Location and source	Oil type and quantity	Species	Impact	Reference
Jan. 1981	San Carlos Creek Jacksonville, Florida; unknown source	Tarlike petroleum substance	West Indian manatee	Male found freshly dead; broken ribs, tarlike substance in cecum and large intestine.	Anonymous (1981a)
Feb. 1981	Jacksonville, Florida; unknown source	Tarlike petroleum substance	West Indian manatee	Body badly decomposed; tarlike substance in cecum and upper large intestine.	Anonymous (1981b)
Mar. 1981	Jacksonville, Florida; unknown source	Tarlike petroleum substance	West Indian manatee	Badly decomposed; plastic and tarlike substance in stomach.	Anonymous (1981c)
Feb. 1982	Little Manatee River, Florida; unknown source	Tarlike petroleum substance	West Indian manatee	Contents of cecum and upper colon were black and smelled like tar.	Anonymous (1982)
Feb. 1983	Persian Gulf; well blowout	Crude oil; est. 30×10^6 gal	Dugong	53 carcasses recovered over a 5-month period.	Anonymous (1983)

to detect it and, if so, avoid it? Anatomic studies suggest that they have fair visual acuity (Cohen *et al.*, 1982), although it can be of little value to them in the murky waters they often inhabit. Their depth perception at close range is apparently poor, and they may bump their head and eyes against objects in the water (Hartman, 1979). They show no reluctance to surface and breathe in the midst of thick mats of floating vegetation (Hartman, 1979), and perhaps would do so even if the surface were covered with oil. They may have a reasonably good sense of smell, judging by the development of the olfactory bulb [(Murie (1872) and Chapman (1875) cited in Ronald *et al.* (1978)]. We need to improve our knowledge of sensory perception in sirenians.

Manatees concentrate their activities in shallow water, often resting at or just below the surface. This behavior would bring them into contact with spilled oil. We might assume that exposure to petroleum would irritate eyes and sensitive mucous membranes; we would expect no unusual harm to epidermis. The manatee's "pelage" is limited to sparse sinus hairs, which may play a role in cutaneous perception. Coating of these structures with oil would not likely result in impairment to a degree greater than that of irritated eyes and lungs.

Judging from their nonselective, generalized feeding habits, manatees might consume tar balls along with their normal food. Despite the many opportunities for manatees to encounter tar, such occurrences have rarely been reported (Anonymous, 1981a,b,c, 1982). From 1974 to 1978, at least 175 dead manatees were examined in Florida waters (Beck *et al.*, 1978; Campbell and Irvine, 1978; Odell and Reynolds, 1979). Human activities, including those associated with motor boats and flood control dams, were frequently cited as causes of death, yet no mention was made of petroleum.

A manatee might also ingest fresh petroleum which some (Geraci and St. Aubin, 1980; Reynolds, 1980) have suggested might interfere with the secretory activity of their unique gastric glands or harm intestinal flora vital to digestion. Whether the long retention time of ingested vegetation in the gut would increase the extent of hydrocarbon absorption is another matter for speculation. Most of the volatile toxic fractions would be assimilated quickly, leaving the high molecular weight compounds to pass in the feces. Perhaps, however, these compounds would be degraded to more absorbable forms by intestinal bacteria, thereby enhancing hydrocarbon uptake.

Bartz and Verinder (1980) evaluated the potential threat of oil development to manatees along the Gulf of Mexico coast of Florida. They predicted no adverse effects, on the presumption that oil spilled offshore would not likely reach critical estuaries and embayments frequented by manatees. While reasonable, this scenario does not preclude the possibility of accidents near shore. The authors voiced greater concern over the projected future increase in vessel traffic, which does have a recognized impact. They and Brownell *et al.* (1978) mention the possibility that spilled oil might affect the quality or availability of aquatic vegetation, but conclude that the effect would be too brief and localized to represent a significant threat.

The need for manatees to occupy small, restricted habitats places them in a vulnerable position, particularly during winter. Oil spills within the confines of preferred river systems and canals would endanger the local population. Manatees able to escape such areas might be forced into colder waters, where thermal stress could complicate the effects of even brief exposure to oil. During summer, displacement would be less detrimental since manatees exploit alternative feeding areas when they deplete local vegetation. This scenario is not one likely to be associated with offshore production or transportation of petroleum. The greater risk is from coastal accidents involving fuel barges or marina storage tanks. For a population whose environment is already under great pressure, even a localized incident could be damaging.

References

Anonymous (1981a). Manatee mortality. *Smithson. Inst., Sci. Event Alert Network (SEAN) Bull.* **6**(1), 26.

Anonymous (1981b). Dead manatee with ingested tar found. *Smithson. Inst., Sci. Event Alert Network (SEAN) Bull.* **6**(2), 31.

Anonymous (1981c). Dead manatee found with ingested tar and plastic. *Smithson. Inst., Sci. Event Alert Network (SEAN) Bull.* **6**(3), 33.

Anonymous (1982). Dead manatee with tar-like substance in caecum and upper colon. *Smithson. Inst., Sci. Event Alert Network (SEAN) Bull.* **7**(3), 37.

Anonymous (1983). Dugongs, other marine life victims of Gulf oil spill. *World Wildl. Fund News* No. 24 (July/Aug.).

Bartz, M. R., and Verinder, S. H. (1980). "OCS Oil and Gas Proposed 1981 Sales A66 and 66: Draft Environmental Impact Statement." Bur. Land Manage., New Orleans Outer Continental Shelf Office, New Orleans, Louisiana.

Beck, C., Bonde, R. K., and Odell, D. K. (1978). Manatee mortality in Florida during 1978. *In* "The West Indian Manatee in Florida" (R. L. Brownell, Jr. and K. Ralls, eds.), pp. 76–85. Florida Dept. of Nat. Resour., Natl. Fish. Wildl. Lab., U.S. Fish Wildl. Serv., Sea World of Florida, Orlando.

Bertram, G. C. L., and Bertram, C. K. R. (1973). The modern Sirenia: Their distribution and status. *Biol. J. Linn. Soc.* **5**, 297–338.

Best, R. C. (1981). Foods and feeding habits of wild and captive Sirenia. *Mamm. Rev.* **11**(1), 3–29.

Best, R. C. (1982). Seasonal breeding in the Amazonian manatee, *Trichechus inunguis* (Mammalia: Sirenia). *Biotropica* **14**(1), 76–78.

Brownell, R. L., Jr., Ralls, K., and Reeves, R. (1978). Report of the West Indian manatee workshop. *In* "The West Indian Manatee in Florida" (R. L. Brownell, Jr. and K. Ralls, eds.), pp. 3–16. Florida Dept. of Nat. Resour., Natl. Fish Wildl. Lab., U.S. Fish Wildl. Serv., Sea World of Florida, Orlando.

Caldwell, D. K., and Caldwell, M. C. (1985). Manatees—*Trichechus manatus, Trichechus senegalensis,* and *Trichechus inunguis. In* "Handbook of Marine Mammals" (S. H. Ridgway and R. Harrison, eds.), Vol. 3, pp. 33–66. Academic Press, London.

Campbell, H. W., and Irvine, A. B. (1978). Manatee mortality during the unusually cold winter of 1976–77. *In* "The West Indian Manatee in Florida" (R. L. Brownell, Jr. and K. Ralls, eds.), pp. 86–91. Florida Dept. of Nat. Resour., Natl. Fish Wildl. Lab., U.S. Fish Wildl. Serv., Sea World of Florida, Orlando.

Cohen, J. L., Tucker, G. S., and Odell, D. K. (1982). The photoreceptors of the West Indian manatee. *J. Morphol.* **173**(2), 197–202.

Domning, D. P. (1982). Evolution of manatees: A speculative history. *J. Paleontol.* **56,** 599–619.

Food and Agriculture Organization (1976). "Mammals in the Seas. Small Cetaceans and Sirenians. Report of the Advisory Committee on Marine Resources Research," ACMRR/MM/SC/3. FAO, Rome.

Gallivan, G. J., and Best, R. C. (1981). Metabolismo e respiracao do peixe-boi da Amazonia (*Trichechus inunguis*). *Acta Amazonica* **11**(4), 679–687.

Gallivan, G. J., Best, R. C., and Kanwisher, J. W. (1983). Temperature regulation in the Amazonian manatee, *Trichechus inunguis. Physiol. Zool.* **56**(2), 255–262.

Geraci, J. R., and St. Aubin, D. J. (1980). Offshore petroleum resource development and marine mammals: A review and research recommendations. *Mar. Fish. Rev.* **42,** 1–12.

Hartman, D. S. (1979). "Ecology and Behavior of the Manatee (*Trichechus manatus*) in Florida," Spec. Publ. No. 5. Am. Soc. Mammal., Washington, D.C.

Irvine, A. B. (1983). Manatee metabolism and its influence on distribution in Florida. *Biol. Conserv.* **25,** 315–334.

Irvine, A. B., Odell, D. K., and Campbell, H. W. (1978). Manatee mortality in the southeastern United States from 1974 through 1977. *In* "The West Indian Manatee in Florida," (R. L. Brownell, Jr. and K. Ralls, eds.), pp. 67–75. Florida Dept. of Nat. Resour., Natl., Fish Wildl. Lab., U.S. Fish Wildl. Serv., Sea World of Florida, Orlando.

Marsh, H., Heinsohn, G. E., and Marsh, L. M. (1984). Breeding cycle, life history and population dynamics of the dugong, *Dugong dugon* (Sirenia: Dugongidae). *Aust. J. Zool.* **32,** 767–788.

Miller, J. (1983). Persian Gulf nations worry as oil spill goes on unchecked. *New York Times,* Aug. 10, A4.

Odell, D. K. (1977). "Age Determination and Biology of the Manatee," Final Rep. Contract No. 14-16-0008-930. U. S. Fish Wildl. Serv., Washington, D.C.

Odell, D. K., and Reynolds, J. E. (1979). Observations on manatee mortality in south Florida. *J. Wildl. Manage.* **43**(2), 573–577.

Powell, J. A., and Waldron, J. C. (1978). The manatee population in Blue Spring, Volusia County, Florida. *In* "The West Indian Manatee in Florida" (R. L. Brownell, Jr. and K. Ralls, eds.), pp. 41–51. Florida Dept. of Nat. Resour., Natl. Fish Wildl. Lab., U.S. Fish Wildl. Serv., Sea World of Florida, Orlando.

Reynolds, J. E., III (1978). Manatees of Blue Lagoon Lake, Miami, Florida: biology and effects of man's activities. *In* "The West Indian Manatee in Florida" (R. L. Brownell, Jr. and K. Ralls, eds.), pp. 25–32. Florida Dept. of Nat. Resour., Natl. Fish Wildl. Lab., U.S. Fish Wildl. Serv., Sea World of Florida, Orlando.

Reynolds, J. E., III (1980). Aspects of the structural and functional anatomy of the gastrointestinal tract of the West Indian manatee, *Trichechus manatus*. Ph.D. Thesis, University of Miami, Coral Gables, Florida.

Reynolds, J. E., III (1981). Behavior patterns in the West Indian manatee, with emphasis on feeding and diving. *Fla. Sci.* **44**(4), 233–241.

Reynolds, J. E., III, and Ferguson, J. C. (1984). Implications of the presence of manatees (*Trichechus manatus*) near the Dry Tortugas Islands. *Fla. Sci.* **47**(3), 187–189.

Ronald, K., Selley, L. J., and Amoroso, E. C. (1978). "Biological Synopsis of the Manatee." Int. Dev. Res. Cent., Ottawa, Ontario, Canada.

Schölander, P. F., and Irving, L. (1941). Experimental investigations on the respiration and diving of the Florida manatee. *J. Cell. Comp. Physiol.* **17,** 169–191.

Shane, S. H. (1984). Manatee use of power plant effluents in Brevard County, Florida. *Fla. Sci.* **47**(3), 180–187.

12

Summary and Conclusions

Joseph R. Geraci and David J. St. Aubin
Department of Pathology
Ontario Veterinary College
University of Guelph
Guelph, Ontario
N1G 2W1 Canada

When oil is drawn from the seabed or transported along oceanic routes, some inevitably escapes. Despite technological advances designed to reduce the chance of inadvertent discharge into the marine environment, there is little hope of eliminating the possibility altogether. The risk of an accident can be calculated from probability tables, but that exercise cannot forecast when and where the next disaster will occur. And as exploration activity increases and expands into new areas, predictions become still less certain. What *is* predictable, however, is that petroleum in the marine environment can have devastating effects on a variety of species. We have focused our attention on marine mammals.

Spilled oil at sea is in a dynamic state. Its fate depends both on its composition and on the environmental conditions at the time of the spill. During the first few hours or days after a spill, low molecular weight fractions quickly evaporate, leaving thick foamy emulsions and heavy residues. These entrain sand and debris, forming tarry aggregates that sink or wash ashore. The slow process of biodegradation eventually recycles the compounds, but in colder waters, this may take decades.

Each major group of marine mammals confronts oil in its own way. Pinnipeds, sea otters, and polar bears are amphibious. Their daily movements may expose them repeatedly to beaches or ice edges where spilled oil accumulates. Their behavior here will determine how and when they will be exposed, and for how long. Some pinnipeds are coastal residents; others are pelagic throughout most of the year, returning to land

Sea Mammals and Oil: Confronting the Risks
Copyright © 1990 by Academic Press, Inc.
All rights of reproduction in any form reserved.

only during brief breeding seasons. Polar bears spend more time on land than in the water, but risk exposure when they cross leads while migrating or pursuing prey. A sea otter spends little time on shore, preferring instead the sanctuary of a kelp bed, which may, in the meantime, have become a trap for oil.

Cetaceans never cross the boundary between sea and land, though some inhabit inshore waters, or are confined to narrow leads through pack ice, where they are likely to encounter spilled oil. Pelagic cetaceans move freely in open water and consequently do not risk prolonged exposure to oil. Sirenians are also entirely aquatic, though they spend most of their time near shore. In winter they seek warm-water refuges. A spill here presents the options of remaining in warm oily water or escaping to clean but dangerously cold seas.

Oil, depending on the form it is in, causes a suite of physiologic and toxic effects. Vapors, and the light fractions that produce them, irritate and damage tissues, especially the sensitive membranes of eyes, mouth, and respiratory tract. Oil compounds are absorbed into the circulation, where they attack liver, nervous system, and blood-forming tissues. Any animal near the site of a fresh spill faces this threat.

The foamy emulsion known as mousse is less toxic, but nonetheless harmful. By sticking tenaciously to vital insulating hairs of sea otters, polar bears, and some species of pinnipeds, like the fur seals, it destroys the animal's ability to maintain thermal balance. Most vulnerable is the sea otter, whose precarious metabolic balance ensures that even casual encounters with oil can be fatal. Oil in cold Arctic waters presents an additional risk to seals, particularly the young. Thick, viscous petroleum can mat the hair, prevent limbs from moving freely, and so affect swimming performance.

On cetaceans, the only comparable surfaces that can collect oil are the haired fringes of the baleen plates. Limited experimental evidence suggests that, at low temperatures, heavy oil can temporarily obstruct the flow of water between the plates and thereby reduce feeding efficiency. This impact is greatest at times of year when baleen whales are feeding intensively to prepare for migration.

There are a few conceivable ways in which a marine mammal might consume raw oil. A mysticete might get a mouthful while skimming the surface; a polar bear might scavenge a contaminated seal carcass, or a seal scavenge an oiled bird; or an otter or bear might lick oil from its fouled fur. Raw oil in this form can be acutely toxic. Moreover, there is the concern that ingesting organisms carrying metabolized residues in their tissues can be unhealthy.

The risk depends on the prey species. Planktonic crustaceans engulf oil droplets during a spill and retain unmetabolized and metabolized petroleum for a week or ten days thereafter. Thus, these organisms are a potential source of contamination to consumers for only a short time after a spill. Similarly, marine fish metabolize assimilated hydrocarbons, excrete them fairly quickly, and thus are not a long-term source of contamination. Bioconcentration through plankton and fish is therefore highly unlikely. In contrast, bivalve molluscs have limited ability to metabolize and excrete oil compounds; they tend to accumulate greater concentrations with time, and through repeated exposure. Molluscs, therefore, pose a lingering threat to walruses, sea otters, and other benthic feeders. For a "scouring" gray whale, the risk of oil ingestion extends beyond tissue residues in food to direct exposure to weathered petroleum in bottom muds.

Irrespective of how they take in these compounds, marine mammals appear to have the liver enzymes required to metabolize and excrete them. This ability limits the accumulation of residues in body tissues and minimizes the probability of residual harm following a spill. This is not to say that the capacity cannot be overwhelmed if an animal swallows large volumes of oil. There is no information from which to assess the effects on an animal feeding in persistently polluted waters, such as those near industrialized or urban areas. Here, it is impossible to distinguish the effects of hydrocarbons from those of the myriad other toxic substances present.

In recent years computer modeling has been used to predict the effect of oil on marine populations. Models have been developed for northern fur seals in the Bering Sea, gray and bowhead whales in Alaskan waters, marine mammals in the Southern California Bight, and sea otters along the California coast. Modeling yields tangible probabilities for oil contact, rates of recovery, and long-term impacts on bioproductivity. However, in building models, broad assumptions must often be made, particularly when data are incomplete. The notable weakness in modeling is that *there is no information on the type and duration of oil exposure required to produce an effect.* For all but the sea otter, the premise that contact is fatal is indefensible; it commonly overestimates the impact of a spill. Models do identify gaps in data, but few gaps that were not already recognized before the models were developed.

How many marine mammals has spilled oil claimed? Few, if any, cetaceans; reports are rare and equivocal. At the other end of the spectrum are sea otters, whose sensitivity to oil was again underscored by the loss of over a thousand individuals after the grounding in 1989 of the *Exxon Valdez*. The vulnerability of pinnipeds lies somewhere in between. Seals have been killed by oil, though large-scale mortality has rarely been observed, even following major disasters near breeding colonies. There have been no reported deaths of polar bears or manatees, perhaps because there has been no major spill where either species congregates.

How will spilled oil influence marine mammals at the population level? Most populations are so large or dispersed as not to be jeopardized by a single spill, whatever the likelihood that an animal or group might be affected or killed. The few exceptions include the small California stock of sea otters, the Mediterranean and Hawaiian stocks of monk seals, and beluga whales in the St. Lawrence River. The greater threat is that oil adds one more perturbation to habitats already deteriorating through abusive commercial fishing and industrial and urban development. We, as stewards of their habitats, should protect the marine mammals from the added burden of oil.

The key to protection is good planning. Decisions on the handling of oil must recognize critical stages in the life histories of the sea animals, such as the times they mate, give birth, or molt, when every effort must be made not to disturb them. Several populations, for example, those of the northern fur seal and Steller's sea lion, are declining for a number of reasons, some of which are surely related to human activities. For these species, the disruption associated with oil spills could be one more hindrance to their recovery. But to schedule oil development and transportation primarily around marine mammals would effectively bring these useful activities to a standstill. Somehow, we must derive a formula allowing the interests of sea mammals and people to coexist. In the foregoing pages, we have tried to assemble the best information toward deriving that formula.

Appendix

Common and scientific names of marine mammal species referred to in text are arranged alphabetically by common name within each family.[1]

Order Carnivora

Family Mustelidae

Sea otter	*Enhydra lutris*

Family Ursidae

Polar bear	*Ursus maritimus*

Family Phocidae (Crawling seals)

Bearded seal	*Erignathus barbatus*
Gray seal	*Halichoerus grypus*
Crabeater seal	*Lobodon carcinophagus*
Harbor seal	*Phoca vitulina*
Harp seal	*Phoca groenlandica*
Hawiian monk seal	*Monachus schauinslandi*
Hooded seal	*Cystophora cristata*
Northern elephant seal	*Mirounga angustirostris*
Mediterranean monk seal	*Monachus monachus*
Leopard seal	*Hydrurga leptonyx*
Ribbon seal	*Phoca fasciata*
Ringed seal	*Phoca hispida*
Saimaa ringed seal	*Phoca hispida saimensis*
Spotted seal	*Phoca largha*
Weddell seal	*Leptonychotes weddelli*

[1] Systematic arrangement from Dale W. Rice, *A List of the Marine Mammals of the World,* 1977, U.S. Department of Commerce, NOAA Technical Report NMFS SSRF-711, and from James H. Honacki *et al., Mammal Species of the World,* 1982, Lawrence, Kansas: Allen Press and the Association of Systematics Collections.

***Family Otariidae* (Walking seals)**

California sea lion	*Zalophus californianus*
Cape fur seal	*Arctocephalus pusillus pusillus*
Guadalupe fur seal	*Arctocephalus galapagoensis*
Northern fur seal	*Callorhinus ursinus*
South American fur seal	*Arctocephalus australis*
Steller's sea lion	*Eumetopias jubatus*
Southern sea lion	*Otaria flavescens*

***Family Odobenidae* (Walruses)**

Atlantic walrus	*Odobenus rosmarus rosmarus*
Pacific walrus	*Odobenus rosmarus divergens*

Order Cetacea

Suborder Mysticeti (Baleen whales)

***Family Balaenidae* (Right whales)**

Bowhead whale	*Balaena mysticetus*
Northern right whale	*Eubalaena glacialis*
Pygmy right whale	*Caperea marginata*
Southern right whale	*Eubalaena australis*

***Family Balaenopteridae* (Rorqual whales)**

Blue whale	*Balaenoptera musculus*
Bryde's whale	*Balaenoptera edeni*
Fin whale	*Balaenoptera physalus*
Humpback whale	*Megaptera novaeangliae*
Minke whale	*Balaenoptera acutorostrata*
Sei whale	*Balaenoptera borealis*

Family Eschrichtiidae

Gray whale	*Eschrichtius robustus*

Suborder Odontoceti (Toothed whales, dolphins, and porpoises)

***Family Monodontidae* (Narwhal and Beluga)**

Beluga	*Delphinapterus leucas*
Narwhal	*Monodon monoceros*

***Family Ziphiidae* (Beaked whales)**

Beaked whales	Various species of the genera *Mesoplodon, Berardius, Hyperoodon, Ziphius* and *Tasmacetus*

Family Delphinidae

Atlantic white-sided dolphin	*Lagenorhynchus acutus*
Bottlenose dolphin	*Tursiops truncatus*
Commerson's dolphin	*Cephalorhynchus commersonii*
Common dolphin	*Delphinus delphis*
Dusky dolphin	*Lagenorhynchus obscurus*
False killer whale	*Pseudorca crassidens*
Hawaiian spinner dolphin	*Stenella longirostris*
Hector's dolphin	*Cephalorhynchus hectori*
Killer whale	*Orcinus orca*
Long-finned pilot whale	*Globicephala melaena*
Northern right whale dolphin	*Lissodelphis borealis*
Pacific white-sided dolphin	*Lagenorhyncus obliquidens*
Risso's dolphin	*Grampus griseus*
Short-finned pilot whale	*Globicephala macrorhynchus*
Spotted dolphins	*Stenella spp.* (Including *S. attenuata, S. dubia, S. frontalis, S. plagiodon*)
Striped dolphin	*Stenella coeruleoalba*
Tucuxi	*Sotalia fluviatilis*
White-beaked dolphin	*Lagenorhynchus albirostris*

***Family Phocoenidae* (Porpoises)**

Dall's porpoise	*Phocoenoides dalli*
Finless porpoise	*Neophocaena phocaenoides*
Harbor porpoise	*Phocoena phocoena*
Vaquita	*Phocoena sinus*

***Family Platanistidae* (River dolphins)**

Chinese river dolphin	*Lipotes uexillifer*

***Family Physeteridae* (Sperm whales)**

Sperm whale	*Physeter catodon*

Order Sirenia

***Family Trichechidae* (Manatees)**

African manatee	*Trichechus senegalensis*
Amazonian manatee	*Trichechus inunguis*
West Indian manatee	*Trichechus manatus*

Family Dugongidae

Dugong	*Dugong dugon*

Author Index

Boldface page entries indicate figures; italic page entries indicate tables.

Subject Index

Boldface page entries indicate figures; italic page entries indicate tables.